RENEWABLE ENERGY: RESEARCH, DEVELOPMENT AND POLICIES SERIES

SOLAR AMERICA: HOW, WHAT AND WHEN?

Renewable Energy: Research, Development and Policies Series

Renewable Fuel Standard Issues
Daniel T. Crowe (Editor)
2009. ISBN: 978-1-60692-289-7

Ethanol and Biofuels: Production, Standards and Potential
Wesley P. Leland (Editor)
2009 ISBN: 978-1-60692-224-8

Ethanol and Biofuels: Production, Standards and Potential
Wesley P. Leland (Editor)
2009 ISBN: 978-1-60876-592-8
(Online Book)

Wind Power: Technology, Economics and Policies
Cedrick N. Osphey (Editor)
2009. ISBN :978-1-60692-323-8

Solar Collectors: Energy Conservation, Design and Applications
Arthur V. Killian (Editor)
2009. ISBN: 978-1-60741-069-0

Solar Collectors: Energy Conservation, Design and Applications
Arthur V. Killian (Editor)
2009. ISBN: 978-1-60876-920-9
(Online book)

Wind Energy in Electricity Markets with High Wind Penetration
Julio Usaola and Edgardo D. Castronuovo
2009. ISBN: 978-1-60741-153-6

Renewable Energy Grid Integration: The Business of Photovoltaics
Marco H. Balderas (Editor)
2009. ISBN: 978-1-60741-324-0

Biomass Gasification: Chemistry, Processes and Applications
Jean-Pierre Badeau and Albrecht Levi (Editors)
2009. ISBN: 978-1-60741-461-2

Renewable Energies: Feasibility, Time and Cost Options
John O'M. Bockris
2009. ISBN: 978-1-60876-006-0

Biological Barriers to Cellulosic Ethanol
Ernest V. Burkheisser
2010. ISBN: 978-1-60692-203-3

Solar America: How, What and When?
Nash M. Perales (Editor)
2010. ISBN: 978-1-60741-333-2

RENEWABLE ENERGY: RESEARCH, DEVELOPMENT AND POLICIES SERIES

SOLAR AMERICA: HOW, WHAT AND WHEN?

NASH M. PERALES
EDITOR

Nova Science Publishers, Inc.
New York

For permission to use material from this book please contact us:
Telephone 631-231-7269; Fax 631-231-8175
Web Site: http://www.novapublishers.com

LIBRARY OF CONGRESS CATALOGING-IN-PUBLICATION DATA

Solar America: how, what and when? / Nash M. Perales.
p. cm.
Includes bibliographical references and index.
ISBN 978-1-60741-333-2 (hardcover : alk. paper)
1. Solar energy--United States. I. Perales, Nash M.
TJ809.95.S563 2009
333.792'30973--dc22

2009035593

Published by Nova Science Publishers, Inc. † New York

CONTENTS

PREFACE

This book is an overview of the Solar Energy Technologies Program in the U.S. The U.S. Department of Energy is committed to "discovering the solutions to power and secure America's future, which include clean energy technologies that not only enhance energy independence but also can reduce greenhouse gas emissions and criteria air pollutants. The Department's robust activities to support the development of solar energy technologies include applied research in support of the Solar America Initiative in the Office of Energy Efficiency and Renewable Energy Solar Energy Technologies Program as well as important exploratory research in the Office of Science.

Chapter 1 - The goal of the Solar America Initiative is to make energy from solar photovoltaics cost competitive with conventional electric power by 2015. Policymakers play a key role in reaching that goal and creating a sustainable environment for the solar industry. The objective of this document is to serve as a guide or template for Solar America Cities to create a comprehensive, city-wide solar plan for their community that facilitates mainstream adoption of solar and serves as a model for other cities to follow.

This guide is broken into discrete areas that the Department of Energy (DOE) has determined to be important components in the constitution of a sustainable solar infrastructure. This guide is a work in progress, and will go through revisions and improvements over the next few years as cities and DOE discover new strategies for mainstreaming solar energy.

Chapter 2 - Energy is essential to our Nation and our lives; we use energy to power homes and businesses and to transport people and goods, making it an important component of our economy. Reliable, clean and affordable energy is critical to energy security and our economy. We can reduce our dependence on foreign sources of energy by diversifying our energy mix using domestic resources.

The U.S. Department of Energy (the Department) is committed to "discovering the solutions to power and secure America's future,"[1] which include clean energy technologies that not only enhance energy independence but also can reduce greenhouse gas emissions and criteria air pollutants. To address the need for increased focus on developing advanced energy technologies and increasing our Nation's focus on building the workforce of tomorrow, in 2006 President Bush launched two important initiatives — the American Competitiveness Initiative (ACI) and the Advanced Energy Initiative (AEI).

Chapter 3 - The inevitable transformation of the electrical grid to a more distributed generation configuration requires solar system capabilities well beyond simple net-metered, grid-connected approaches. Time-of-use and peak-demand rate structures will require more sophisticated systems designs that integrate energy management and/or energy storage into the system architecture. Controlling power flow into and from the utility grid will be required to ensure grid reliability and power quality. Alternative protection strategies will also be required to accommodate large numbers of distributed energy sources.

This document provides an overview of the R&D needs and describes some pathways to promising solutions. The solutions will, in many cases, require R&D of new components, innovative inverter/controllers, energy management systems, innovative energy storage and a suite of advanced control algorithms, technical methodologies, protocols and the associated communications. It is expected that these solutions will help to push the "advanced integrated system" and "smart grid" evolutionary processes forward in a faster but focused manner.

Chapter 4 – Hearing before the Committee on Energy and Natural Resources, United States Senate one hundred tenth Congress Second Session to Consider the Value and Examine the Progress of Electricity Generation from Concentrating Solar Power

Chapter 5 - On Monday, March 17, 2008 the House Committee on Science & Technology, Subcommittee on Energy and Environment will hold a hearing entitled, *''Utility-Scale Solar Power: Opportunities and Obstacles,''* at the Pima County Administration Building Hearing Room, Tucson, Arizona.

The Subcommittee's hearing will explore the potential for utility-scale solar power to provide a significant fraction of U.S. electric generating capacity and the challenges to achieving this goal. The specific technologies to be discussed include solar thermal technology, concentrating photovoltaics and distributed solar power. Transmission, regulatory and financial issues will also be examined, along with a look at the government and private industry roles in the development of utility-scale solar power—and enabling productive partnerships between them.

In: Solar America: How, What and When?
Editor: Nash M. Perales
ISBN: 978-1-60741-333-2

Chapter 1

SOLAR IMPLEMENTATION GUIDE FOR CITIES

Solar America Cities

INTRODUCTION

The goal of the Solar America Initiative is to make energy from solar photovoltaics cost competitive with conventional electric power by 2015. Policymakers play a key role in reaching that goal and creating a sustainable environment for the solar industry. The objective of this document is to serve as a guide or template for Solar America Cities to create a comprehensive, city-wide solar plan for their community that facilitates mainstream adoption of solar and serves as a model for other cities to follow.

This guide is broken into discrete areas that the Department of Energy (DOE) has determined to be important components in the constitution of a sustainable solar infrastructure. This guide is a work in progress, and will go through revisions and improvements over the next few years as cities and DOE discover new strategies for mainstreaming solar energy.

Each of the sections in this plan is organized in the following manner

- **Description:** A short description of the incentive, policy, or guideline
- **Providers:** Best positioned entities to act upon or provide this incentive or policy
- **Best Practices:** Describes how different cities and states have actually implemented these policies and incentives to stimulate the marketplace

RULES, REGULATIONS, AND POLICIES

Net Metering & Interconnection Policies

Description

For customers who generate their own electricity, *net metering* allows for the flow of electricity both to and from the customer, typically through a single, bi-directional meter. With net metering, during times when a customer's generation exceeds the customer's use, electricity from the customer flows back to the grid, offsetting electricity consumed by the customer at a different time. In effect, the customer uses excess generation to offset electricity that the customer otherwise would have to purchase from the utility at the utility's retail rate. Net metering is required by state law or regulations in 42 U.S. states and the District of Columbia, although some of these laws and regulations only apply to investor-owned utilities. Several utilities — mostly municipal utilities — offer net metering voluntarily. There is no federal net-metering requirement, and there has been little federal guidance on this issue.

The term *interconnection* refers to the technical and procedural process by which a customer connects an electric-generating system to the power grid. In general, interconnection standards specify the technical, contractual, metering and rate issues by which system owners and utilities must abide.

The Public Utility Regulatory Policies Act of 1978 (PURPA) requires investor-owned utilities and certain other utilities with annual retail sales of at least 500 million kilowatt-hours (kWh) to interconnect with customer-generators, but this law has not been fully effective at the state level. The Federal Energy Regulatory Commission (FERC) has adopted interconnection standards for generators up to 20 megawatts (MW) in capacity, but these standards generally apply only to transmission-level interconnection.

Providers

Usually states implement net-metering rules; however, cities with municipal utilities may also implement net-metering rules.

Interconnection standards for distribution-level interconnection are generally set forth at the state level, although standards adopted by many states apply only to investor-owned utilities. Some municipal utilities have developed their own interconnection standards for net-metered systems.

Best Practices and Examples

The most successful net-metering rules and programs include the following provisions:

- The maximum individual system capacity should be at least 2 MW.
- All utilities — including municipal utilities and electric cooperatives — should be required to offer net metering.
- All customer classes should be eligible.
- The limit on the aggregate capacity of all net-metered systems in a utility's service territory should be at least 5% of the utility's annual peak load.
- Any customer net excess generation (NEG) at the end of a billing cycle should be credited to the customer's next bill at the utility's full retail rate until the customer leaves the utility's system. Alternatively, a utility should compensate the customer for NEG at the utility's avoided-cost rate at the end of a 12-month period.
- Fair, safe and reasonable interconnection standards should be adopted for net-metered systems.
- Utilities should not be permitted to impose application fees on prospective net-metering customers.
- Utilities should not be permitted to impose any special charges or fees for net-metering customers, or force customers to switch to a different tariff.

- Customers should retain ownership of all renewable-energy credits (RECs) associated with the customer's generation.

New Jersey's net-metering rules are widely considered to be the best in the United States. New Jersey's rules incorporate most of the best practices listed above, with one notable exception — the rules only apply to investor-owned electric utilities and not to municipal utilities or electric cooperatives.[1] In combination with an aggressive Renewables Portfolio Standard (RPS) and solar set-aside, generous financial incentives, excellent interconnection standards and a favorable regulatory climate, New Jersey's net-metering rules have led to the deployment of more than 2,000 photovoltaic (PV) systems since 2004. Colorado's rules and Maryland's rules are very similar to New Jersey's.

Of the dozen or so municipal utilities that offer net metering, Lakeland Electric has adopted one of the most pro-solar programs. Lakeland Electric allows commercial PV systems up to 500 kilowatts (kW) in capacity to net meter, while most other municipal utilities that offer net metering voluntarily limit individual systems to 10 kW.[2] A handful of electric cooperatives voluntarily offer net metering; these programs are typically limited to systems with a 10-kW maximum capacity.

Map of net metering availability:
http://www.dsireusa.org/documents/SummaryMaps/ NetMetering Map.ppt

The most successful state-level interconnection standards include the following provisions:

- The maximum individual system capacity should be at least 10 MW.
- Standards should include three or four separate levels of review to accommodate systems based on system capacity, complexity and level of certification. Small, certified systems should be processed as quickly as possible, while larger, uncertified systems should require closer review. Larger systems that do not export electricity should require a less rigorous review process than larger systems that do export electricity.
- Application costs should be kept to a minimum, particularly for smaller systems.
- Reasonable, punctual procedural timelines should be adopted and enforced.
- A standard form agreement should be used.
- An external disconnect switch should not be required.
- Clear, transparent technical screens should be established.
- Interconnection to area networks should be permitted, with reasonable limitations where appropriate.
- Utilities should not be permitted to require customers to purchase liability insurance.
- Utilities should not be permitted to require customers to indemnify utilities, absent a reciprocal provision.

Roughly two dozen states have adopted interconnection standards for distributed generation. Colorado's interconnection standards, which are similar to the federal interconnection standards set forth in FERC Order 2006, incorporate most of the provisions listed above. New Jersey's interconnection standards also are among the best in the United

States, although New Jersey's standards only apply to systems with a maximum capacity of 2 MW. At the local level, few municipal utilities and electric cooperatives have voluntarily developed guidelines for interconnection. (Notable exceptions are those municipal utilities and electric cooperatives that have voluntarily developed guidelines for net metering.)[3]

For more detailed information about interconnection and net metering issues, please refer to the Connecting to the Grid guide, published by the North Carolina Solar Center and the Interstate Renewable Energy Council

(http://www.dsireusa.org/documents/Policy Publications/IC_Guide.pdf).

Public Benefits Funds

Description

Public benefits funds (sometimes referred to as clean energy funds) are typically state-level programs developed to assure continued support for renewable energy, energy efficiency, and low-income programs. Several such funds were established through the electric utility restructuring processes of the late 1990s.

Providers

States generally provide public benefits funds by means of taxing ratepayers a small amount in their utility bill.

Best Practices and Examples

Most of the 17 existing state public benefits funds are supported through a surcharge on utility customer electricity consumption (e.g., 0.2 cents/kWh) but have also been established as a result of utility merger settlements. Renewable energy funds are used to provide direct incentives and financing for renewable energy projects, business development activities, research and development, and education programs. These funds have been instrumental in spurring the growth of solar markets in recent years.

Municipalities that have authority over their local utilities may also consider establishing funds via a dedicated surcharge or flat monthly fee to support solar programs. Municipal utilities in California, for example, have been administering solar programs supported by public benefits funds as directed by state policy for a number of years. Lakeland Electric, a municipal utility in Florida, recently authorized a flat $0.25/month public benefits charge for all customers to support its solar water heating and conservation programs.[4]

Map of states with public benefits funds for renewable energy:
http://www.dsireusa.org/documents/SummaryMaps/PBF Map.ppt

Solar Access Laws

Description

Solar access laws are designed to protect a consumer's right to install and operate solar energy systems on a home or business, including access to sunlight (also referred to as solar access).

Providers

Typically states pass legislation related to solar access laws; however, local governments can pass ordinances that may effectively have the same result.

Best Practices and Examples

About a dozen states have passed laws that limit or prohibit restrictions that neighborhood covenants and/or local ordinances can impose on the use of solar equipment. For example, Arizona law protects property rights to solar access by dissolving any local covenant, restriction, or condition attached to a property deed, contract, or security agreement that restricts the installation or use of solar energy devices; however, property deeds, contracts, or security agreements that were entered into prior to April 17, 1980 are exempt.[5]

Some of the key elements to include in solar access laws include providing a clear and quantifiable standard for what constitutes an unreasonable restriction on solar energy systems (i.e., changes for aesthetic reasons cannot increase installation costs by more than 10%); defining the types of structures (residential, commercial, etc.) covered by the law; and awarding costs and reasonable attorney fees to the prevailing party in any civil action arising from disputes with homeowners associations.

An even more common type of solar access law at the state level is the solar easement. Easements allow for the rights to existing access to solar energy on the part of one property owner to be secured from an owner whose property could be developed in such a way as to restrict that resource. This easement is transferred with the property title. While this approach offers some protection for consumers, the policy does not address potential barriers imposed by local governments or homeowner associations. California's Solar Rights Act and Solar Shade Control Act combine elements of solar easements and protections against government and neighborhood restrictions.

At the local level, communities use a variety of different mechanisms to protect solar access, including solar access ordinances, development guidelines requiring proper street orientation, and zoning ordinances that contain building height restrictions to avoid shading neighboring solar panels.

Building Energy Standards That Mandate Solar Use

Description

Regulations may be passed at the state and local level requiring building standards that exceed national code and meet more aggressive energy efficiency standards such as those contained in IECC 2006 and ASHRAE 90.1 2006. If allowable by the state, cities may choose to go one step further and incorporate solar energy mandates into local building codes and standards.

Maps and overviews of the status of commercial and residential state energy codes can be found at: http://www.bcap-energy.org/map_page.php

Providers

State, city, and county governments may have the authority to require such standards.

Best Practices and Examples

Municipalities and state governments can play a critical role in supporting renewable energy by buying electricity from renewable resources, developing solar and green building design standards, or committing to installing a specified level of solar capacity on public buildings. About 10 states and dozens of cities require that a certain percentage of electricity purchased for government buildings come from renewable resources. In these cases, solar energy is not typically a big part of the renewable electricity mix.

Solar energy systems can be used to help meet green building certification requirements. Although a handful of states specifically target solar installations, these policies generally only require "consideration" of solar. In general, policies that merely encourage integrating solar into new buildings when feasible or that require consideration of solar on public buildings is not as effective as specific capacity or investment requirements.

Oregon, for example, requires that all new state public building projects and major renovations invest in solar technologies at a level of at least 1.5% of the total contract price.[6] California's statute calling for solar energy equipment to be installed on all state buildings and state parking facilities by 2007, where feasible, as well as a more recent *Green Building Initiative* to reduce grid-based energy use by 20% at major state-owned facilities, has led to the development of four megawatts of solar capacity on state buildings.' As a local example, San Francisco established a solar-specific capacity objective of 50 MW on public buildings by 2012 and has a bond revenue mechanism in place to help finance those projects.

Solar Set-Asides in Renewables Portfolio Standards Policies

Description

Renewables Portfolio Standards (RPS) require that a certain percentage of a utility's retail energy sales or new generating capacity be derived from renewable resources, e.g., 10% of electric sales must be from renewable energy by the year 2020. The state or municipality implementing an RPS may choose to require a certain percentage of the RPS to be met specifically with solar energy.

Providers

Because states typically implement RPS policies, cities should work with the states in which they reside to encourage them to adopt such policies. Municipalities that have authority over their local utilities may also choose to adopt RPS policies to promote renewable energy development.

Best Practices and Examples

More than two dozen states have implemented such renewable generation mandates. Although wind and biomass are the predominant resources used to satisfy RPS obligations, a growing number of states are incorporating a "carve-out" or "set-aside" within the RPS stipulating that a portion of the renewable energy percentage or overall retail sales be derived from *solar* resources. For example, New Jersey, Maryland, and Delaware have each set aggressive targets for 2% of the state's electricity mix to be generated from solar resources, which together could result in more than 3,000 MW of solar capacity over the next 15 years. Eight other states and the District of Columbia have adopted solar or distributed generation set-asides as part of their RPS policies as wel1.[8]

Best practices for promoting solar through RPS policies involve:

- Establish an explicit solar set-aside in the RPS that ramps up over time.
- Develop a mechanism for tracking, verifying and trading solar renewable energy certificates (SRECs).
- Impose a monetary penalty or include an alternative compliance payment provision for electricity suppliers that do not meet solar generation requirements.
- Require long-term power-purchase or SREC contracts to ensure project developers can access financing.
- Encourage small-scale, distributed systems.

Colorado's Renewable Energy Standard exemplifies some of these key elements. For investor-owned utilities (IOUs), the requirement begins at 3% of retail electricity sales in 2007, rising incrementally to 20% by 2020. At least 4% of the renewable energy must be generated by solar-electric technologies and at least one-half of the solar share must be generated by solar-electric systems located on-site at customers' facilities. SRECs are tradable. IOUs must offer a solar rebate of at least $2 per watt for systems up to 100 kW and can offer an up-front payment for SRECs. Utilities must develop standardized SREC contracts with a minimum term of 20 years. Electric cooperatives and municipal utilities are subject to a lower renewables standard, and there is no solar requirement. However, solar electricity generated by a facility that begins operation before July 1, 2015, receives 300% credit for RPS-compliance purposes.[9]

Municipalities that have authority over their local utilities may also choose to adopt RPS policies to promote renewable energy development. Cities leading the way in this regard include Columbia (MO), Austin (TX), and Fort Collins (CO). For example, the Austin City Council adopted a resolution for its municipal utility, Austin Energy, to meet 30% of all energy needs through the use of renewable resources by 2020, including at least 100 MW of solar power.[10]

Map of states with solar provisions in RPS policies:
http://www.dsireusa.org/documents/ummaryMaps/solar

Financial Incentives

Direct Incentives

Description

Direct incentives are used to 'buy down' the cost of solar systems and come in several forms including grants, rebates, and performance-based incentives. These incentives, typically covering 20% to 60% of project costs and ranging from a few hundred to millions of dollars, have played a significant role in encouraging solar installations. Rebates are typically disbursed to customers once the project is up and running and are typically awarded on a $/W basis. Performance-based incentives, on the other hand, provide project owners with cash payments based on electricity production on a $/kWh basis over a specified duration. Hybrid approaches — upfront rebates based on *expected* performance — have also been developed; these incentives are based on capacity ($/W) but take into consideration system rating, location, tilt and orientation, and shading to adjust the incentive. Payments based on performance or expected performance rather than capital investments are gaining prominence among program administrators as a way to maximize system design and installation.

Providers

The majority of direct incentives for solar projects are implemented at the state level and by publicly-owned utilities. However, there are examples of cities that provide direct incentives through municipally-owned utilities as well.

Best Practices and Examples

- Offer a generous incentive level with stable, long-term funding that decreases over time as the market matures. A good example is the California Solar Initiative — a 10-year, $3+ billion program that provides incentives for solar installations in all sectors. Incentive levels will automatically be reduced over the duration of the program in 10 steps based on the aggregate capacity of solar installed. In this way, incentive reductions are linked to levels of solar demand rather than an arbitrary timetable.[11]
- Establish a consistent but cost-effective quality-assurance mechanism to protect consumers and guarantee adequate system performance. Many state programs employ one or more of the following provisions: equipment requirements, design guidelines, pre-approved or certified installers, or incentives based on actual or expected performance.
- Create a feed-in tariff that supports solar at either the city or state level. Michigan Assemblywoman Kathleen Law introduced a bill in September 2007 that creates the first comprehensive renewable energy feed-in tariff (FIT) introduced into any U.S. legislature.[12]
- Incorporate an installer training and development strategy into incentive programs. The New York State Energy Research and Development Authority's (NYSERDA) efforts in solar workforce development are exemplary. NYSERDA provides numerous training and marketing assistance opportunities for solar installers and has invested nearly $1,000,000 in developing seven accredited solar training centers and continuing education programs across the state.[13]
- Consider offering higher incentive levels for high-value solar applications. Massachusetts, for example, gives a bonus incentive for systems incorporated into high-performance or "green" buildings and homes; systems that use components manufactured in-state; and installations for affordable housing developments or public buildings.[14] Wisconsin gives a higher incentive level for solar equipment installed by contractors with a specialized solar credential — certification by the North American Board of Certified Energy Practitioners (NABCEP).[15]
- Design an easy and concise application process.
- Allow flexibility for program modifications.
- Develop a coordinated package of policies to complement direct incentives, including net metering, low-interesting financing, and tax incentives.
- Foster utility support and cooperation to ensure a quick and easy grid-interconnection process for photovoltaic systems.

- Work with other state agencies and relevant stakeholder groups to educate the public about renewable energy technologies and to market the incentive program.
- Track the details of program use, costs, and energy savings/production to enable program evaluation and improvement.

Low-Interest Loan Programs

Description

Low-interest financing for the purchase of solar energy equipment is a mechanism used to reduce the burden of high, up-front costs of solar equipment by providing consumers loans at a low interest rate. Repayment schedules vary and are usually determined on an individual project basis, but some offer a repayment term of up to 7-10 years.

Providers

States, utilities, and municipalities have used low interest loans to encourage the installation of solar energy equipment. This mechanism is used in about a dozen states primarily to support projects for the non-residential sector. Utility and local loan programs typically target the residential sector. Municipal or county programs may consider partnering with a local bank or community economic development organization to secure favorable terms or to structure interest rate buy-downs provided by the municipality.

Best Practices and Examples

- Aspen, Colorado's Community Office for Resource Efficiency (CORE), for example, has partnered with the Community Bank of Colorado to provide financing for photovoltaic or solar hot water systems. The typical loan term is five years and CORE pays the interest rate, resulting in a zero-interest loan for consumers.[16]
- The City of Ashland, Oregon's Conservation Division offers a solar water heating program to residential electric customers who currently use an electric water heater. Under "The Bright Way to Heat Water Program," qualified home owners may take advantage of the City's zero-interest loan program or a cash rebate. Customers choosing a loan repay it as part of their monthly utility bill. Interested customers are provided site evaluations, consumer education, information about available solar systems, and names of qualified contractors.[17]
- New Jersey's Clean Energy Program, administered through the New Jersey Board of Public Utilities, has offered local governments and schools a low-interest, longterm financing program to combine energy efficiency and renewable energy incentive programs Financing under this program covered the entire incremental cost of energy efficiency and renewable energy projects, enabling the governmental entities to finance the projects completely, with no up-front capital necessary.[18]
- Low-interest loans for renewable-energy systems are available to residential and small business customers of all Wisconsin Public Power, Inc. (WPPI) member utilities. This incentive is available for solar water-heating systems, solar space-heating systems for buildings that use electricity as the primary heat source,

photovoltaic (PV) systems up to 20 kilowatts (kW) in capacity, wind-energy systems up to 20 kW in capacity, and repairs to existing systems. Customers can borrow from $2,500 to $20,000, at an interest rate of 1.99%. Loan terms vary from three to 10 years Financing is through Fannie Mae's Residential Home Energy Improvement Program and underwritten by Energy Finance Solutions (EFS), a service of Wisconsin Energy Conservation Corporation. Eligibility is based on the customer's credit score, bankruptcy history and debt-to-income ratio.[19]

Income/Investment Tax Credits

Description

Tax credits are directly deducted from the tax owed by the owner of the solar system. They typically range from 10% to 50% of project costs and can serve as an important driver to promote solar deployment in states where public benefits funds or other direct funding sources are not available.

Providers

More than a dozen states offer personal and/or corporate tax credits to help offset the expense of purchasing and installing solar energy equipment.

Best Practices and Examples

One of the weaknesses often attributed to tax incentive policies is that entities without a tax liability, such as government facilities, non-profits, and schools, are not eligible for the incentive despite their increasing interest in employing solar technologies. However, some states have developed provisions to extend tax credit benefits to non-taxed entities. Under Oregon's business energy tax credit "pass-through" option, for example, a project owner may transfer a tax credit to a pass-through partner in return for a lump-sum cash payment (the net present value of the tax credit) upon completion of the project.[20] This mechanism allows non-profit organizations, schools, governmental agencies, and other public entities and businesses with *or* without a tax liability to use the credit by transferring their tax credit to a partner with a tax liability.

Arizona's solar tax credit allows a third-party company that finances and installs a solar energy system on a tax-exempt organization's facility to claim the credit, which results in lower overall project costs for the organization.[21] This type of arrangement in which a third party owns and operates a system on a public building and sells the electricity through a power-purchase agreement is gaining in popularity. Extending tax benefits to applications on public buildings can benefit municipalities seeking to install solar on their own facilities.

States have broadened tax credit programs in other ways to encourage a greater level of solar adoption. For example, while most tax credit programs target project owners, a few states, including Oregon, Utah, and Rhode Island, allow homebuilders who install solar energy systems to claim the credit in an effort to encourage the construction industry to integrate solar into new developments. Installing solar during building construction rather than as a retrofit improves the economics of such projects.

Although solar tax credits are typically state-level policies, municipal governments that impose income, franchise or other similar taxes can consider credits or exemptions to encourage solar adoption. New York City, for example, is considering offering a credit against its municipal income tax for commercial solar energy systems, as these projects are not eligible for the New York state solar tax credit.

Property Tax Financing Districts

Description

While state and federal subsidies have made it more cost effective for many residential and commercial property owners to install solar electric and solar thermal systems, high up-front costs still present a barrier to more widespread solar adoption. Property tax financing for solar installations, as has been proposed by the City of Berkeley, could enable more property owners to install solar energy systems by allowing them to pay for the system through a long-term assessment on their property tax bill.

Best Practices and Examples

In November 2007, the Berkeley City Council approved a proposal by Mayor Tom Bates to make Berkeley the first city in the nation that would allow property owners (residential and commercial) to pay for energy efficiency improvements and solar system installations as a voluntary long-term assessment on their individual property tax bill. The City would provide the funding for the project from a bond or loan fund that it repays through assessments on participating property owners' tax bills for 20 years. The program is now in development, with a June 2008 target date for the launch of the pilot phase of the project.

The Financing District would solve many of the fmancial hurdles facing property owners. First, there would be little upfront cost to the property owner. Second, the upfront costs would be repaid through a voluntary tax on the property, therefore funding approval would not determined directly by a property owner's credit or the equity in the property. Third, the total cost of the solar system and energy improvements would be comparable to financing through a traditional equity line or mortgage refinancing because the well-secured bond would provide lower interest rates than are commercially available. Fourth, the tax assessment would be transferable between owners. Therefore, if a property is sold prior to the end of the 20-year repayment period, the next owner would take over the assessment as part of their property tax bill.

Property Tax Incentives

Description

Property tax incentives for solar installations typically follow a simple model that excludes the added value of solar energy equipment in the valuation of the property for taxation purposes.

Providers

States or local governments may impose property taxes and incentives to offset them.

Best Practices and Examples

Nearly half the states in the United States offer this incentive, but in some cases, such as New Hampshire, Virginia, and Colorado, state law allows local governments to provide exemptions from property taxation but does not require it.[22] In states where local governments have the authority to offer such exemptions, municipalities can use that authority to insulate residents and businesses that choose to install solar energy systems from higher property taxes.

Sales Tax Incentives

Description

Sales tax incentives for solar projects, implemented in more than a dozen states, usually take the form of an exemption from the state sales tax for the cost of solar energy equipment.

Providers

Typically states provide for sales tax exemption, although some cities impose sales taxes as well.

Best Practices and Examples

In some states, the exemption is restricted to a particular sector, e.g., residential, or to systems that meet certain size requirements. Ideally, such exemptions would apply to all solar energy installations. As with property tax exemptions, local governments that have the authority to offer exemptions from local sales taxes may choose to offer this added benefit to residents and businesses that purchase and install solar energy systems.

Expedited Permitting and Fee Waivers for Solar Projects

Description

Municipalities may consider other forms of assistance such as waiving building permits, plan check, or other fees consumers would normally face when installing a solar energy system. “Top-of-the-stack” or fast-track permitting, which often translates into savings, is another benefit local governments can provide to consumers or developers who install solar energy systems.

Best Practices and Examples

The City of Tucson, for example, will credit a building permit applicant for the permit fee up to a maximum of $1,000 for the installation of a qualifying solar energy system. The city allocated a total of $100,000 worth of credits for new buildings and an additional $100,000 worth of credits for existing buildings.[23] Marin County, California, in addition to offering a small rebate for solar energy systems, also waives design reviews for most solar installations for existing buildings if flush-mounted and offers over-the-counter permits for most solar installations on existing buildings.[24]

Industry Development Incentives

Description

Incentives can be used as a recruitment tool by cities and states to attract solar businesses to locate in a particular area. By providing these financial incentives, cities and states are able to attract the burgeoning solar industry to areas within the state where other forms of manufacturing may have once played a key role. This not only leads to job creation but also to positive public relations. Incentives can be used not only to promote the establishment or expansion of manufacturing operations, but also to support research, development and commercialization activities; partnerships with private venture capital funds to invest in clean energy companies; and business development activities for distributors and installers.

The most common industry incentives offered are loans, grants, tax abatements, tax credits, and tax exemptions, or a commitment by the state to purchase a set amount of the product. A number of states currently offer incentive-based programs for industry development (CA, CO, HI, MA, MI, MT, NC, NE, NM, NY, OR, TX, VA, WA, and WI). Many of the loan and grant programs are run through state-managed renewable energy funds, some of which are receiving funding through state legislative appropriations.

Best Practices and Examples

- CO's New Energy Economic Development (NEED) Fund: In 2007, the Colorado Legislature appropriated money to the Governor's Energy Office (GEO) from the limited gaming tax and the severance tax to establish and maintain the state's Clean Energy Fund. NEED, funded through the Clean Energy Fund, provides grants, loans and other financial incentives to attract manufacturers of renewable energy to the state. A total of $350,000 was available during the first funding round. The second round of funding, totaling about $650,000, will be available during the next competitive cycle with preference given to proposals that demonstrate the ability to provide matching funds.
- MA's Business Expansion Initiative: The Massachusetts Technology Collaborative (MTC), which administers the state's Renewable Energy Trust Fund, offers loans to support renewable energy companies entering or expanding the manufacturing stage of commercial development. Loan amounts range from $500,000 to $3,000,000, and are available for up to 50% of capital expenses and related spending over a two-year period.
- NY's Renewable, Clean Energy, and Energy Efficient Product Manufacturing and Incentive Program: Administered by the New York State Energy Research and Development Authority (NYSERDA) and funded through the New York System Benefits Charge (SBC), this incentive program goal is to increase the manufacturing of renewable, clean, and energy-efficient products in the state by providing funds to manufacturers that wish to develop or expand facilities producing eligible products. The funding is provided in milestone phases with the limit of $1.5 million per project. Phase I provides money, up $75,000 with a cost share of 50% required, for facility and site characterization activities. Phase II covers pre-production development funding, no more than 20% of the total fund

and up to a maximum of $300,000. This phase also requires a 50% cost share. Phase III is a production incentive payment based on the sale of clean energy products produced. The remaining 75% of total funding is available for this phase, subject to a 75% cost share and a total funding limit of $1.5 million per project.

TECHNICAL TRAINING

Training for Installers and Code Officials

Description

As markets develop within cities, the need for experienced solar installers becomes vital. Solar installers serve an essential role in ensuring quality and maintaining the legitimacy of cities' solar programs, particularly in nascent solar markets where the acceptance of solar is not necessarily prevalent. Even when using the finest equipment and system designs, solar energy systems can still fail if they are improperly installed. Equally important are qualified code inspectors that can ensure that all solar energy systems are installed in a safe manner

Providers

Many different organizations may provide technical training if they have the expertise and facilities to provide it adequately. Training has been offered in the past by a variety of organizations including community colleges, non-profit organizations, universities, and trade organizations.

Best Practices and Examples

- The New York State Energy Research and Development Authority's (NYSERDA) efforts in solar workforce development are exemplary. NYSERDA provides numerous training and marketing assistance opportunities for solar installers and has invested nearly $1,000,000 in developing seven accredited solar training centers and continuing education programs across the state.[25]
- Codes and Standards are the backbone of the success of the Solar America Initiative (SAI). DOE selected New Mexico State University and its partners to lead a new Solar Codes and Standards Working Group. The Working Group will address code development and outreach activities in areas of critical importance to solar market penetration, such as interconnection procedures, net metering, product safety, and international standards coordination. This five-year effort will create a major improvement in the responsiveness, effectiveness, and accessibility of codes and standards to U.S. solar stakeholders at all levels, including cities.
- The North American Board of Certified Energy Practitioners (NABCEP) offers PV and solar thermal installer certification programs.[26]
- The Florida Solar Energy Center, the North Carolina Solar Center, and New Mexico State University have developed courses for installers and code officials to serve both state and national needs for training.[27]

OUTREACH

Outreach

Description

In order for solar energy to maximize its potential share of the energy market, the public must be educated on the benefits and challenges associated with its development and deployment. Solar outreach efforts can be tailored to target a number of audiences, including consumers (homeowners and businesses), utilities, financial institutions, educators and students, policy makers and regulators, skilled labor force, and potential retailers of solar technologies (e.g. big box retailers).

Lack of communication, information dissemination, and consumer awareness will limit solar's potential in our energy market. Outreach efforts in the form of media campaigns, education and training programs, and high-visibility events, competitions, and demonstration projects are just a few examples of outreach activities that can be implemented at the city level.

Providers

Potential providers of solar outreach include state and local governments, community organizations, colleges and universities, non-profit organizations, utilities, and industry associations.

Best Practices and Examples

A number of cities and states have solar outreach programs in place. These include:

- Portland, Oregon's *Solar Now!* campaign[28] is a one-year effort to grow demand for solar hot-water and photovoltaic systems. Within its first year, the project aims to get 100 systems installed in Portland in high-visibility locations and by high- visibility businesses and community members. The campaign is produced by the City of Portland Office of Sustainable Development in collaboration with Energy Trust of Oregon, Oregon Department of Energy, and Solar Oregon.
- In partnership with the Focus on Energy Program, WisconSUN[29] (Wisconsin's solar energy initiative) provides resources and programs uniquely designed to meet the needs of Wisconsin's solar energy community As part of its marketing efforts, WisconSUN produces consumer-friendly information on solar energy systems specific to Wisconsin and easy access to the solar energy marketplace for site owners and consumers. Its Web site includes case studies and white papers on solar energy systems, links to funding sources, and links to dealers, installers, architects and engineers in the state.
- Pacific Gas & Electric's (PG&E) Solar Schools Program[30] is teaching the value of alternative energy by turning school buildings in California into hands-on science experiments. This award-winning program is making science fun and teaching students how their everyday actions can impact the environment. This year, PG&E will donate up to $2 5 million to support bright ideas grants, demonstration solar installations and training courses for teachers.

End Notes

[1] http://www.dsireusa.org/library/includes/incentive2.cfm?Incentive Code=NJO3R&state=NJ&CurrentPagel D=1
[2] http://www.dsireusa.org/library/includes/incentive2.cfm?Incentive Code=FL15R&state=FL&CurrentPagel D=1& RE=1 &EE=0
[3] http://www.dsireusa.org/library/includesitype.cfm?EE=1&RE=1
[4] www.ases.org/solar2007/presentations/wednesday/200pm/forums/3-solar therma1/3-curry.pdf
[5] http://www.azleg.state.az.us/FormatDocumentasp?inDoc=iars/33/00439.htm&Title=33&DocType=ARS
[6] http://www.newniles.orgielectricity/solaror.html, http://www.energy
[7] http://www.newrules.org/electricity
[8] http://dsireusa.org/library/includesitype.cfm?EE=0&RE=1
[9] http://dsireusa.org/library/includes/incentivesearch.cfm?Incentive Code=CO24R&Search=Type &type=RP S & CurrentPagelD=2&EE=O&RE=1
[10] http://dsireusa.org/library/includes/incentivesearch.cfm?Incentive_Code=TX11R&Search=Type&type=RP S & CurrentPagelD=2&EE=O&RE=1
[11] http://dsireusa.org/library/includes/incentive2.cfm?Incentive Code=CA134F&state=CA&CurrentPagelD= 1 &RE =1 &EE=1
[12] http://www.renewableenergyaccess.com/rea/news/storyjsessionid=8608C1FA3F9BB6CEDE146D22360A D046? id=50004
[13] http://text.nyserda.org/programs/Energy Resources/photovoltaics.asp#training
[14] http://dsireusa.org/library/includes/incentive2.cfm?Incentive Code=MA22F&state=MA&CurrentPagelD=1 &RE =1&EE=O
[15] http://dsireusa.org/library/includes/incentive2.cfm?Incentive_Code=WIl4F&state=WI&CurrentPageID=1 &RE =1 &EE=O
[16] http://www.aspencore.org/sitepages/pid77.php
[17] http://dsireusa.org/library/includes/incentive2.cfm?Incentive Code=0R1OF&state=0R&CurrentPageID=1 &RE =1 &EE=O
[18] http://www.nj cleanenergy. com/
[19] http://www.dsireusa.org/library/includes/incentive2.cfm?Incentive Code=WIl3F&state=WI&CurrentPageID=1 &RE=1 &EE=1
[20] http://www.dsireusa.org/library/includes/incentive2.cfm?Incentive Code=0R03F&state=0R&CurrentPageID=1 &RE=1&EE=1
[21] http://www.dsireusa.org/library/includes/map2.cfm?CurrentPageID=1 &State=AZ&RE=1&EE=1
[22] http://dsireusa.org/library/includesitype.cfm?EE=0&RE=1
[23] http://dsireusa.org/library/includes/incentive2.cfm?Incentive Code=AZ26F&state=AZ&CurrentPagelD=1&RE =1 &EE=1
[24] http://dsireusa.org/library/includes/incentive2.cfm?Incentive Code=CAO8F&state=CA&CurrentPagelD=1&RE =1 &EE=1
[25] http://text.nyserda.org/programs/Energy Resources/photovoltaics.asp#training
[26] http://www.nabcep.org/pv installer.cfrn
[27] http://www.fsec.uc£edu/en/education/cont ed/pv/igcpvs/installers.php, http://www.ncsc.ncsu.edu/, http://www nmsu edu/%7Etdi/Photovoltaics/Codes-Stds/Codes-Stds.html
[28] http://www.portlandonline.corn
[29] http://www.wisconsun.org/
[30] http://www.pge.com/solarschools/

In: Solar America: How, What and When?
Editor: Nash M. Perales

ISBN: 978-1-60741-333-2

Chapter 2

SOLAR AMERICA INITIATIVE: A PLAN FOR THE INTEGRATED RESEARCH, DEVELOPMENT, AND MARKET TRANSFORMATION OF SOLAR ENERGY TECHNOLOGIES

U.S. Department of Energy, Energy Efficiency and Renewable Energy

1.0 INTRODUCTION

Energy is essential to our Nation and our lives; we use energy to power homes and businesses and to transport people and goods, making it an important component of our economy. Reliable, clean and affordable energy is critical to energy security and our economy. We can reduce our dependence on foreign sources of energy by diversifying our energy mix using domestic resources.

The U.S. Department of Energy (the Department) is committed to "discovering the solutions to power and secure America's future,"[1] which include clean energy technologies that not only enhance energy independence but also can reduce greenhouse gas emissions and criteria air pollutants. To address the need for increased focus on developing advanced energy technologies and increasing our Nation's focus on building the workforce of tomorrow, in 2006 President Bush launched two important initiatives — the American Competitiveness Initiative (ACI) and the Advanced Energy Initiative (AEI).

Department of Energy Releases Its Strategic Plan

In 2006, the Department released its DOE Strategic Plan, which outlines DOE goals. These included the following strategic goals, which the Department's solar-related activities address.

- **Energy Strategic Goal:** Promoting America's energy security through reliable, clean, and affordable energy.
  - Goal 1.1. Energy Diversity:
    - Increase our energy options and reduce dependence on oil, thereby reducing vulnerability to disruption and increasing the flexibility of the market to meet U.S. needs
  - Goal 1.2 Environmental Impacts of Energy:
    - Improve the quality of the environment by reducing greenhouse gas emissions and environmental impacts to land, water, and air from energy production and use
  - Goal 1.3 Energy Infrastructure:
    - Create a more flexible, more reliable, and higher capacity U.S. energy infrastructure
- **Science Strategic Goal:** Strengthening U.S. scientific discovery, economic competitiveness, and improving quality of life through innovations in science and technology.
  - Goal 3.1 Scientific breakthroughs:
    - Achieve the major scientific discoveries that will drive U.S. competitiveness, inspire America, and revolutionize approaches to the Nation's energy, national security, and environmental quality challenges
  - Goal 3.2 Foundations of Science:
    - Deliver the scientific facilities, train the next generation of scientists and engineers, and provide the laboratory capabilities and infrastructure required for U.S. scientific primacy.
  - Goal 3.3 Research Integration:
    - Integrate basic and applied research to accelerate innovation and to create transformational solutions for energy and other U.S. needs.

One of the renewable resources of focus in the AEI is solar energy, and in support of the development of solar energy technologies, the President in 2006 launched the Solar America Initiative (SAI). Solar energy is a clean, abundant, widespread and renewable energy source that can be used to increase electricity generating capacity. Electricity produced from solar energy will not only reduce demand for natural gas — which is increasingly imported — but will also reduce the greenhouse gas emissions from traditional combustion-driven electricity generation. Various technologies can capture solar energy and convert it into other useful forms of energy, such as electricity and heat.

The Department's robust activities to support the development of solar energy technologies include applied research in support of the Solar America Initiative in the Office of Energy Efficiency and Renewable Energy Solar Energy Technologies Program as well as important exploratory research in the Office of Science. The President's Solar America

Initiative supports the goals of the DOE 2006 Strategic Plan, which identifies the strategic themes of energy security and scientific discovery and innovation, both of which are key drivers of the DOE solar activities. SAI also aligns with the solar-related research and development provisions and broad energy goals outlined in the National Energy Policy and the Energy Policy Act of 2005 (EPACT 2005). In EPACT 2005, Congress expressed strong support for decreasing dependence on foreign energy sources and the cost of renewable energy generation and delivery. The support of Congress, as well as incentives and support from state governments, will be critical to achieving the SAI goals.

Advanced Energy Initiative

The Advanced Energy Initiative provides for a 22% increase in funding for clean-energy technology research at the U.S. Department of Energy. The AEI is an integrated plan to improve America's energy independence by changing how we power our homes and offices by generating more electricity from clean coal, advanced nuclear power, and renewable resources such as solar and wind, and by developing more efficient vehicles that are powered by electricity, biofuels and hydrogen.

Advanced Energy Initiative Goals — Changing How We Power Our Homes and Businesses

- Complete the President's commitment to $2 billion in clean coal technology research funding, and move the resulting innovations into the marketplace.
- Develop a new Global Nuclear Energy Partnership (GNEP) to address spent nuclear fuel, eliminate proliferation risks, and expand the promise of clean, reliable, and affordable nuclear energy.
- Reduce the cost of solar photovoltaic technologies so that they become cost-competitive by 2015, and expand access to wind energy through technology.

Advanced Energy Initiative Goals — Fueling Our Vehicles

- Develop advanced battery technologies that allow a plug-in hybrid-electric vehicle to have a 40-mile range operating solely on battery charge
- Foster the breakthrough technologies needed to make cellulosic ethanol cost-competitive with corn-based ethanol by 2012.
- Accelerate progress towards the President's goal of enabling large numbers of Americans to choose hydrogen fuel cell vehicles by 2020.

This plan describes the Solar America Initiative and details the efforts through 2015 that the Department will undertake to spur widespread commercialization, acceptance and adoption of clean solar technologies across the United States.

The Department has developed a coordinated strategy for accomplishing the goals of the Solar America Initiative which will include:

- Conducting accelerated research and development to improve the materials performance and reduce the cost of advanced photovoltaic (PV) systems
- Developing new manufacturing technology to lower process costs and increase throughput for enabling expanded U.S. manufacturing capability
- Providing technical expertise and building stakeholder consensus to resolve regulatory, institutional, infrastructure, and education-related barriers to technology acceptance
- Accelerating deployment of new solar technologies through promoting demonstrations and early adopter activities consistent with the Energy Policy Act of 2005
- Supporting the demonstration and deployment of energy technologies through collaborative efforts with the private sector and public sector entities

ENERGY POLICY ACT OF 2005

In July of 2005, Congress passed the first comprehensive energy legislation in over a decade. The Energy Policy Act of 2005 (P.L. No: 109-058) was signed into law by President Bush on August 8, 2005. The law supports many of the objectives outlined by President Bush in the launch of the Advanced Energy Initiative— strengthening our nation's energy security and infrastructure; reducing our dependence on foreign oil; and expanding the use of clean renewable energy. Title IX. subtitle C of the bill focuses on renewable energy research, development and demonstration. Specifically, section 931(A) states that " [t]he Secretary shall conduct a program of research, development, demonstration, and commercial application for solar energy" to accomplish the high-level objectives supported by Congress.

The Solar America Initiative at a Glance

- The SAI will boost research and development (R&D) on the manufacturing processes and product designs that have the best chance of making PV systems less expensive, more efficient, and highly reliable.
- The President's fiscal year (FY)2007 request of $148.4 million for SAI-related activities represents a more than $66 million increase over FY 2006 funding for solar energy R&D.
    - The FY 2007 budget request for PV-related activities was $139.8 million.
    - The FY 2007 budget request for concentrating solar power (CSP)-related activities was $8.9 million, representing a 20% increase over the FY 2006 appropriation.
- The Department will conduct research and development to reduce cost and expand production of PV by funding industry-led teams collaborating across the value chain.These projects will expand U.S. annual domestic PV manufacturing capacity and also focus on reducing the cost of electricity from PV.
- The Department will also support early-stage companies as they take promising prototype PV cells from laboratory benches into commercial pilot production, leveraging DOE funding and technical assistance from the National Renewable

Energy Laboratory (NREL) and Sandia to bring the next generation of PV technologies to market after 2011.

- The Department will also address non-technological barriers to widespread deployment of solar technologies such as codes, standards, certification, and technical training.
  - The Department will establish partnerships with States and utilities to catalyze collaboration in the design of regulations and incentives that promote adoption of solar technologies.
  - Solar "showcase"projects will demonstrate novel large-scale market applications of PV, thereby retiring risk and allowing future private financing of similar projects.
  - The Department will work with city governments that are combining regulation, training, and other measures on route to becoming desirable locations for establishing solar businesses and marketing solar products.
  - The Department will establish Energy Star labeling for solar hot water heaters to ensure product performance.
- The Department will facilitate cost reductions in CSP systems by addressing three factors: further technology development, volume production, and scale-up in plant or project size.

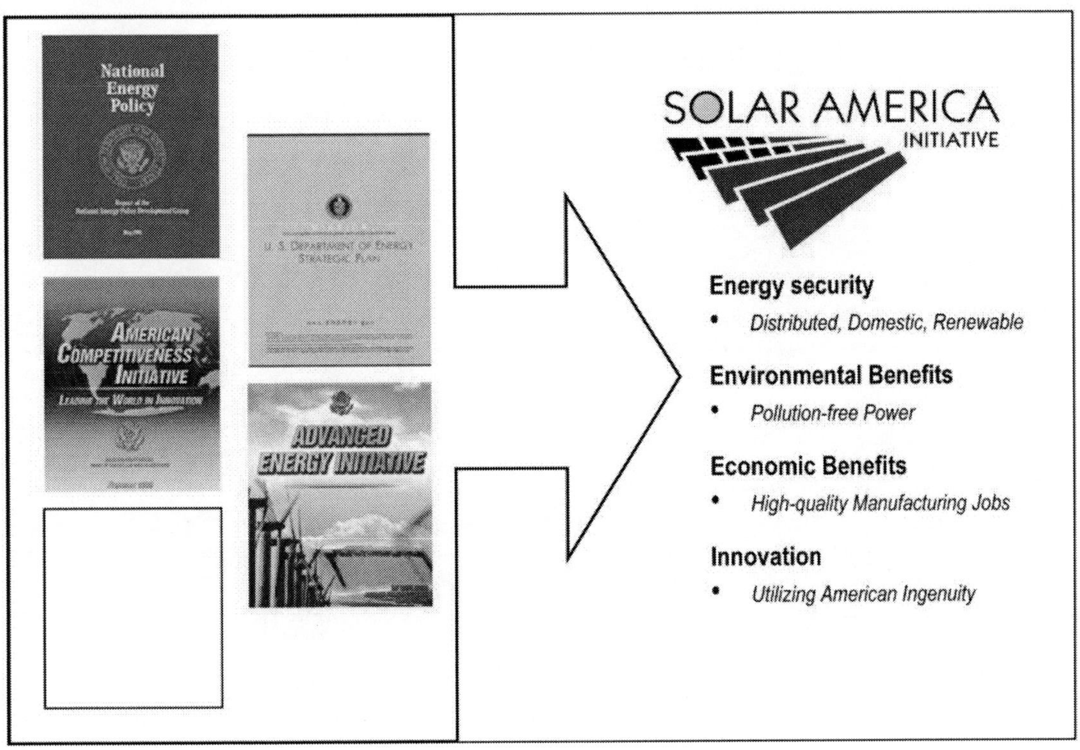

Accomplishing the goals of AEI, and more specifically SAI, will require greater understanding of solar energy technologies than can be gained through basic science research. The "use- inspired" research being sponsored by the DOE Office of Science (OS) may lead to some of the breakthroughs needed for the SAI to be successful, as well as revolutionary concepts that will be useful beyond the 2015 timeframe. The applied research in the Office of Energy Efficiency and Renewable Energy (EERE) includes furthering technology advancements in performance, cost, and reliability, scaling up of manufacturing facilities, reducing infrastructure barriers to deploying the technology, and innovative partnerships between government, industry, national laboratories, non-profit organizations, and universities.

SAI Targets

In support of the SAI, the Department has established performance and cost stretch targets for PV systems in three target markets: grid-connected residential, commercial, and utility applications. As the SAI stretch targets are met, significant market penetrations are forecast to occur across all market segments. The cost of electricity is assessed by the kilowatt-hour, so DOE's targets for PV and CSP systems are based on the *Levelized Cost of Energy* (LCOE) delivered by these systems. LCOE is a measure of total costs of a system (over its expected lifetime) divided by the expected energy output (over its useful lifetime), with appropriate adjustments for time, value of money, etc. The overall cost goals for SAI are shown in Figures 1-1 and 1-2. These targets are based on Energy Information Administration (EIA) projections of relatively flat electricity prices (in real terms)

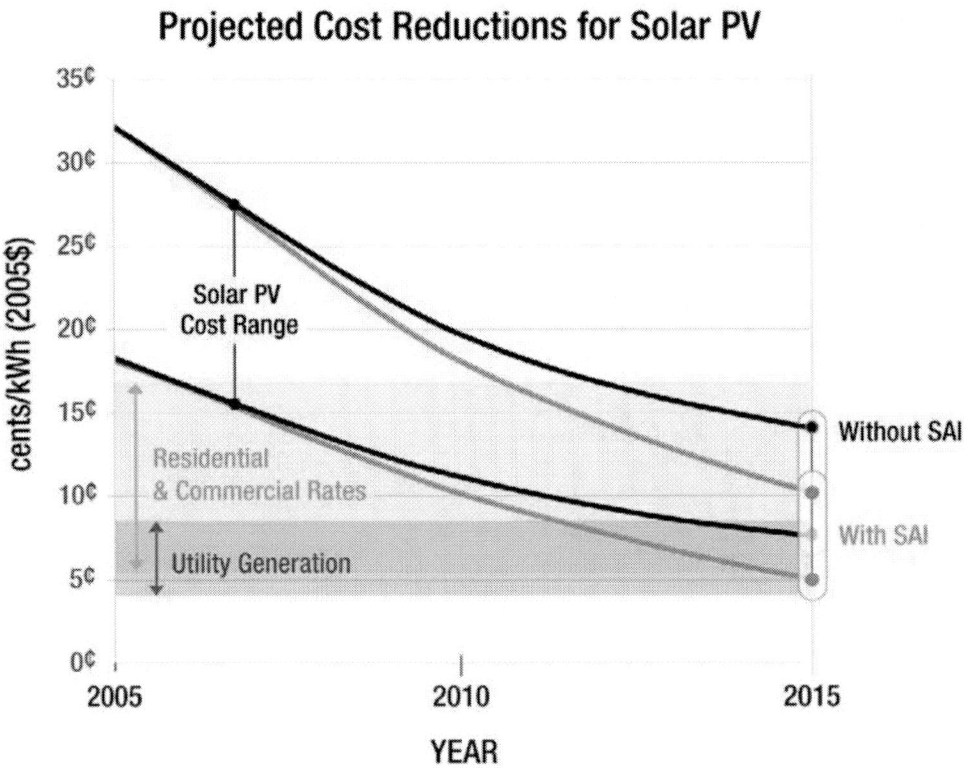

Figure 1-1. Projected PV Cost Reductions across All Market Sectors within SAI

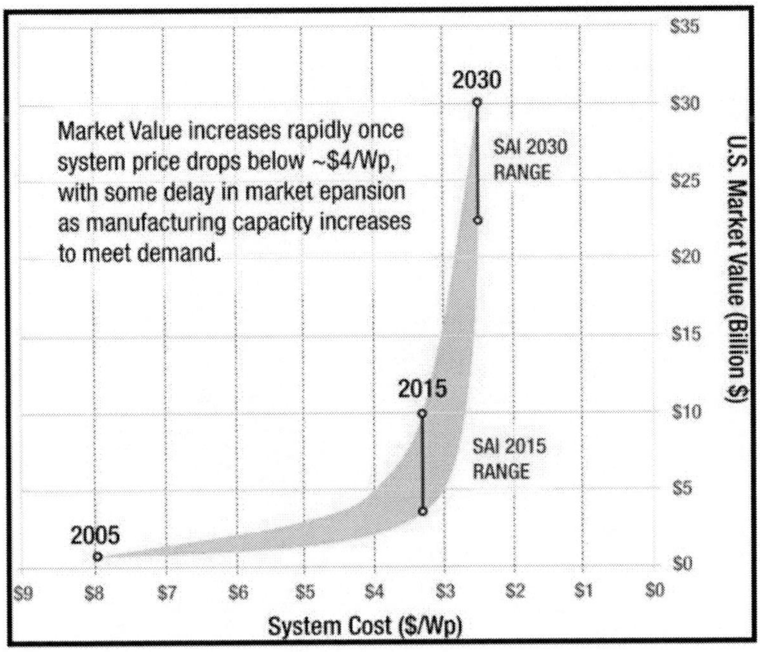

Figure 1-2. Projected Market Value for PV Systems, 2005-2030

The Benefits of the SAI

Displacing a significant amount of conventional electricity production with new energy technologies is a major challenge. The U.S. electric generation system is enormous, with roughly 1,000 GW of generating capacity currently in place. In contrast, in 2005, the total U.S. installed PV capacity was 0.44 GW,[3] i.e., less than 0.1% of total U.S. generating capacity. Yet, solar energy is available in all regions of the country and can provide significant amounts of energy in places like New York and Minnesota, not just places like Texas and California. In fact, according to National Renewable Energy Laboratory projections, if every single-family home in America had a 3 kilowatt (kW) photovoltaic (PV) system on its roof, these combined homes could generate more than 420 billion kilowatt-hours (kWh) of electricity-more than 35% of the entire residential electricity demand for the United States.[4]

Solar energy technologies can improve energy independence by supplementing current electricity generation capacity. In recent years, natural gas combined-cycle power plants have made up the majority of new generating capacity. As a result, demand for natural gas in the power sector has increased steadily over the past 15 years, even in the face of dramatically higher prices. The tight balance between natural gas supply and demand has led to a more volatile market, which can respond dramatically to weather-related events and geopolitical developments. At present, 85% of U.S. natural gas demand is met through domestic production.[5] Natural gas consumption, however, is projected to grow to 74 billion cubic feet per day by 2025,[6] up from 61 billion cubic feet per day in 2004.[7] Diversifying our energy mix with solar technologies can help to mitigate the effects of fluctuating natural gas prices.

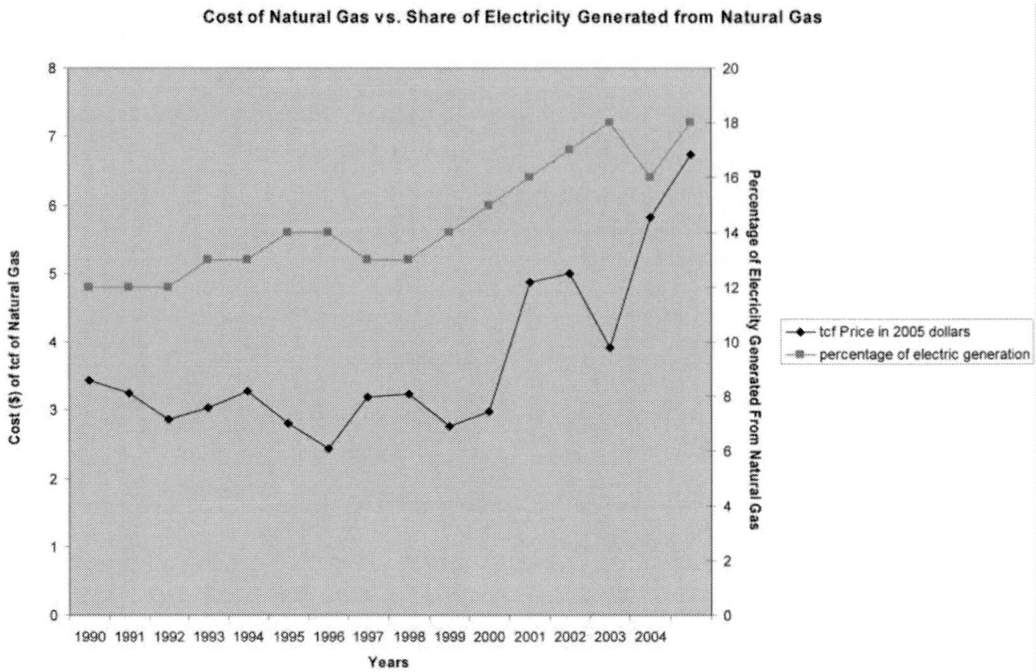

Figure 1-3. Cost of Natural Gas vs. Share of Electricity Generated from Natural Gas

Under the SAI, distributed PV can play a significant role in displacing the need for new electricity generation capacity. Assuming that for every MW of PV installed, 0.6 MW of new capacity is displaced (a reasonable average for the United States), various scenarios show that PV could displace 10-25% of new capacity additions in 2015, and up to 40% of new capacity additions in 2030. Projections for baseline, moderate, and high *cumulative* capacity additions scenarios by 2030 are depicted in the following figure.[8]

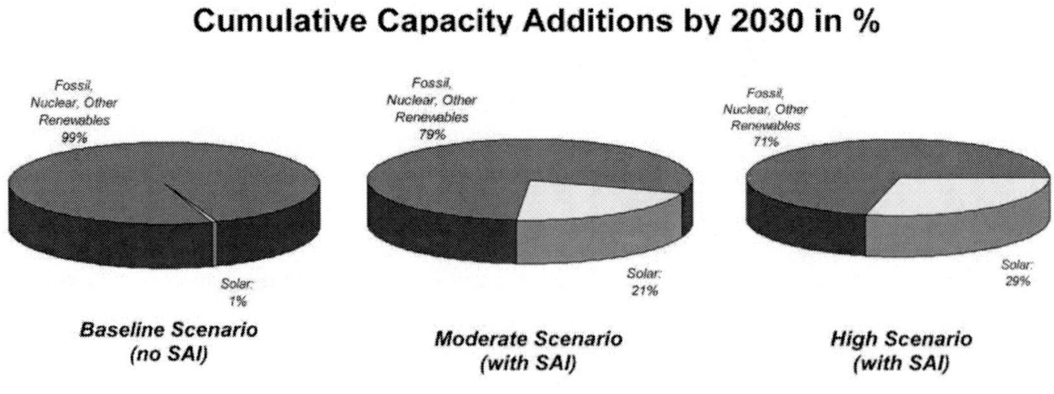

Figure 1-4. Solar as Part of New Capacity Additions 2030 (Baseline, Low and High Scenario)

In addition to benefiting our National Security by increasing energy independence, solar energy also serves to make the electric grid more secure and less susceptible to terrorist attack. Use of decentralized photovoltaics will enhance the reliability of our aging electric grid by reducing the demand placed on it, as well as reducing the impact of a potential terrorist attack on our centralized electricity infrastructure.

Electricity produced from solar energy will also reduce greenhouse gas emissions in the United States. Electricity generation using natural gas combined-cycle power plants or conventional coal combustion technology generates both criteria pollutant and greenhouse gas emissions. Many electricity generating processes also require significant quantities of water for cooling. Solar electricity, by contrast, emits no pollutants or greenhouse gases into the air, and requires very little water. Using solar energy technologies in the place of conventional methods to produce 100MW of electricity would result in a reduction of 191,000 tons/yr of CO_2, 7.4 tons/yr of NOx, and 4.5 tons/yr of CO.[9] Based on this data, DOE estimates that the deployment of new solar technology installations under SAI will triple the amount of carbon emissions avoided by 2015.

Investment in the solar industry through SAI will also ensure that the U.S. solar industry is able to capture an increased market share of the anticipated global and domestic demand for solar technology, supporting more jobs for skilled workers in manufacturing, production, and installation. Under the SAI, direct PV-related employment is projected to result in the creation of 10,000-30,000 jobs by 2015 and 67,000-89,000 jobs by 2030.[10]

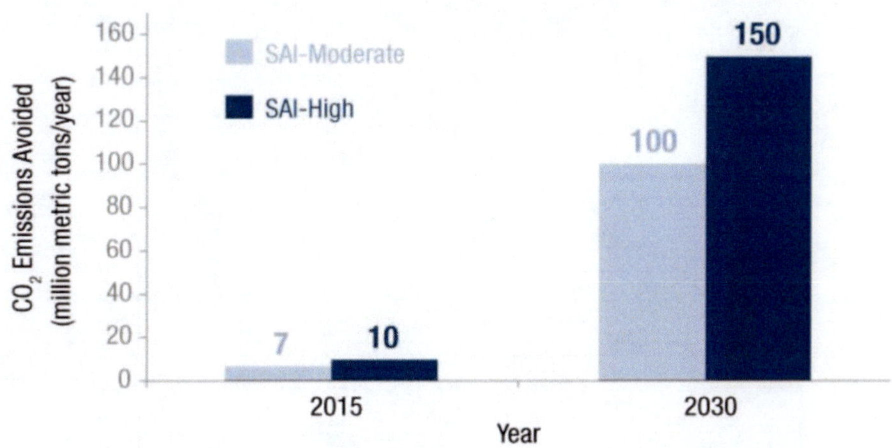

Figure 1-5. The SAI accelerates greenhouse gas reductions by avoiding CO_2 emissions. By 2030 the SAI is projected to avoid 100-150 million more metric tons of CO_2 annually than the baseline.

2.0 Markets

PV has often been the power source of choice for remote applications, based on cost and proven reliability. However, lower costs and recent advances in PV efficiencies have enabled the PV market to broaden rapidly to include utility, distributed generation, and building-integrated applications. A major milestone in the evolution of PV markets occurred in 2002 when sales for grid-connected PV applications decisively surpassed sales for remote applications.[11]

- The PV industry has been expanding rapidly during the last decade. Much of this growth has been driven by PV-targeted subsidies in Germany, Japan, and a number of U.S. states (e.g., California, Arizona, and New Jersey).
- Global PV production increased from about 60 MW in 1994 to almost 1.8GW in 2005. 2006 production is estimated to be 2400 MW.
- During this period, the most rapidly growing PV markets were for grid-connected PV systems installed on residential and commercial buildings.
- Although solar is one of the fastest-growing high-tech industries, the U.S. solar industry faces strong and growing challenges from competitors in Europe and Asia.
- After decades of global leadership in PV market share, the United States has fallen behind other countries in the manufacturing of PV (see Figure 2-1), such as Japan and Germany China, Korea, and India are now also investing heavily in PV research and development to solidify their respective market positions.
- Similarly, the United States' lead in Concentrating Solar Power (CSP) technology is being challenged by Germany, Spain, and Israel[12] These and other countries have captured growing global market share through dedicated federal investments in technology R&D and market incentives.[13]

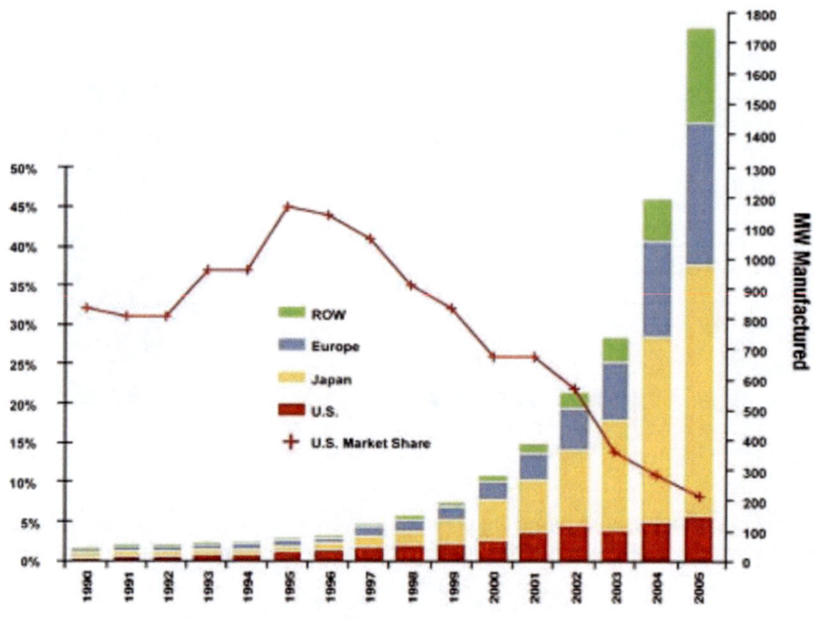

Figure 2-1. Global PV Market Share by Country

The global PV market is expected to continue its rapid growth through the next decade (see Figure 2-2), and grid-tied residential and commercial markets are expected to be the primary drivers of this growth. According to the recent U.S. PV Industry Roadmap, the domestic PV industry is expected to parallel the global growth of the global PV industry during this time with California leading the way, followed by New Jersey and other states that have aggressive solar programs.[14]

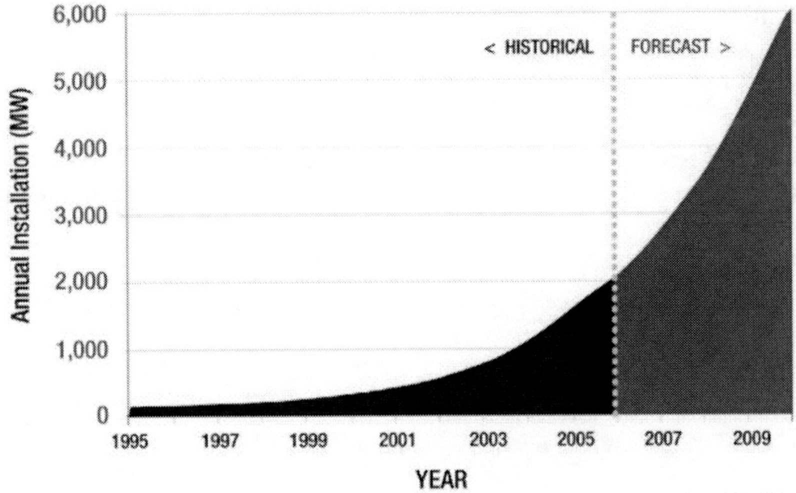

Figure 2-2. Global PV Installation Forecast

The U.S. Department of Energy will address these shortfalls of the U.S. PV industry using the market-driven approach of the SAI, minimizing the time between laboratory developments and product deployment. This will facilitate cost reductions at a rate that otherwise could not be achieved.

One consequence of this rapid growth has been the emergence of a supply shortage of solar- grade silicon. Solar-grade silicon is a key input for crystalline PV cells/modules, the dominant PV technology in the marketplace today. This supply shortage, which is most likely temporary with new supplies coming on line between 2007 and 2009, has created an opportunity for thin- film PV and concentrator technologies, which use little-to-no polysilicon feedstock, to accelerate their move from pilot production into manufacturing and large-scale production.[15]

The silicon shortage coupled with the high demand for solar products will likely keep PV prices higher over the next two years as demand continues to outpace supply.[16] The SAI is targeting manufacturing and process improvements that will lower the production cost for PV technologies. Ultimately, prices will fall to reflect the lower production cost as PV production capacity and silicon feedstock production increase to meet demand.

While the U.S. has fallen behind in the PV market, technology improvements and various State- funded incentives are stimulating domestic demand for solar systems. Although the prospects for continued growth in the global PV industry are significant, many factors could influence how rapidly U.S. markets expand.

Federal and State PV Incentives

- There is a need to significantly reduce the cost of PV systems via research to improve performance and lower materials, manufacturing, and installation costs.
- Increases in market opportunities will be realized through the reduction of institutional barriers, including the lack of interconnection standards and net-metering provisions for distributed energy

- **Residential Solar and Fuel Cell Tax Credit:** The Energy Policy Act of 2005 establishes a 30% tax credit capped at $2,000 for the purchase and installation of residential PV property.

- **Commercial Solar and Fuel Cell Tax Credit:** EPAct 2005 also establishes a 30% tax credit for the purchase and installation of PV systems on commercial property (no cap).

In addition, several States have solar incentives in place. Among California's many solar programs are:

- **California Solar Initiative:** The goal of this $2.8 billion, 10-year program is to create 3,000 megawatts of new solar capacity by 2016.

- **Property Tax Exemptions for Solar Systems:** This is a property tax exemption for certain types of solar energy systems installed by December 31, 2009.

- **Zero Energy New Home Program:** This program develops new home designs that will optimize energy efficiency and onsite solar generation to decrease household energy bills.

New Jersey also has many solar programs, including:

- **Clean Energy Financing for Local Schools and Governments:** The New Jersey Board of Public Utilities offers local governments and schools a low-interest, long-term financing program to combine energy efficiency and renewable energy incentive programs.

- **Solar and Wind Energy Systems Exemption:** New Jersey offers a full exemption from the State's 6% sales tax for all solar and wind-energy equipment.

The Concentrating Solar Power Market and SAI Potential

Concentrating Solar Power technologies are most often applied in centralized power production. Although power from CSP currently costs more than other renewable options such as wind, utilities are becoming more aware of the potential economic benefits of CSP deployment. There are several reasons for utility interest in CSP. CSP power production aligns closely with their periods of peak demand. The problems of solar intermittency can be overcome with thermal storage or hybridization with natural gas, allowing plants to dispatch power to the line when it is needed. In some regions, such as the Southwest, the widespread availability of solar energy provides flexibility in locating CSP power plants near existing or planned transmission lines.

Parabolic trough technologies have had the most commercial success in the CSP market so far.

- Solar Electric Generating Stations (SEGS) projects in California have reached a capacity of 354 MW.
- The first SEGS plant was completed in 1985, and all nine plants continue operating today.
- There has also been a recent renewal of commercial CSP activity in the US and Spain. A 1 MW trough plant began operation in Arizona in 2006 and a 64 MW plant is scheduled to begin operation in Nevada in 2007.[17]
- The cost of power from a new CSP trough plant built today is estimated to be 12-140 ¢ /kWh.[18]

Significant progress has been made on reducing both component costs and O&M expenses associated with trough plants in the last several years. For example, advances in thermal storage, using molten salt have made it possible to provide the dispatchable power desired by the electric power industry. This ability to store solar energy makes the technology particularly attractive to utilities because it gives them the option of using the power when they need it most.

Since the late 1970's *parabolic dish-engine* technologies have seen several demonstrations and pre-commercial deployments, but there are currently no commercial dish systems in operation. A prototype six-dish, 150 kW power plant built with private funds is now operating at the National Solar Thermal Test Facility at Sandia National Laboratories. The experience gained from the prototype plant has been helpful in reducing the capital cost of these systems, and the operational experience will improve reliability and reduce O&M costs.

Markets for CSP are being driven by new policy incentives and technology improvements, and this is resulting in renewed worldwide market interest. Favorable power purchase agreements are leading to commercial projects in Spain, and European suppliers are in competition with American suppliers for these markets.[19] Recent market activity suggests that large deployments of dish-Stirling systems will soon be a reality in the U.S. as well.

- On August 10, 2005, Stirling Energy Systems (SES) and Southern California Edison announced a 20-year power purchase agreement that will result in the construction of a 4,500 acre, 20,000 dish plant in the Mojave Desert. The plant is slated to have an output of 500MW, and the agreement provides for possible expansion to 800MW.[20]
- San Diego Gas & Electric announced that it had also signed an agreement with SES, and intends to purchase all of the output from a 300MW, 12,000 dish plant for a period of 20-years. The deal contains provisions for the plant, which is to be located in Southern California's Imperial Valley, to be expanded by as much as 600MW in the future.[21]

There are a number of factors that determine the cost of power from CSP plants, such as the intensity of solar insolation, interest rates, and the cost of commodity materials such as glass and steel. The size of the plant is also important, with plants greater than 200 MW often providing the lowest cost power.[22]

SAI will facilitate cost reductions in CSP by addressing three factors: further technology development, volume production, and scale-up in plant or project size. Technology development includes evolution in the performance and reliability of specific technology components (receivers, concentrators, reflectors, and balance of solar field), improvements in construction techniques and reductions in O&M costs due to learning experience as more projects are installed. Volume production brings significant cost reductions due to decreases in manufacturing cost, material procurement costs, standardized engineering, and project development costs. Large power plant sizes or multiple plants in a single project invoke economies of scale in equipment and systems.

An example of the expected cost reduction in CSP is illustrated in Figures 2-3 and 2-4. Estimates are given for 2015 deployment levels up to 4 GW. This represents a development and deployment plan for the relatively mature parabolic trough technology, which is also a reasonable scenario for the other CSP technologies.

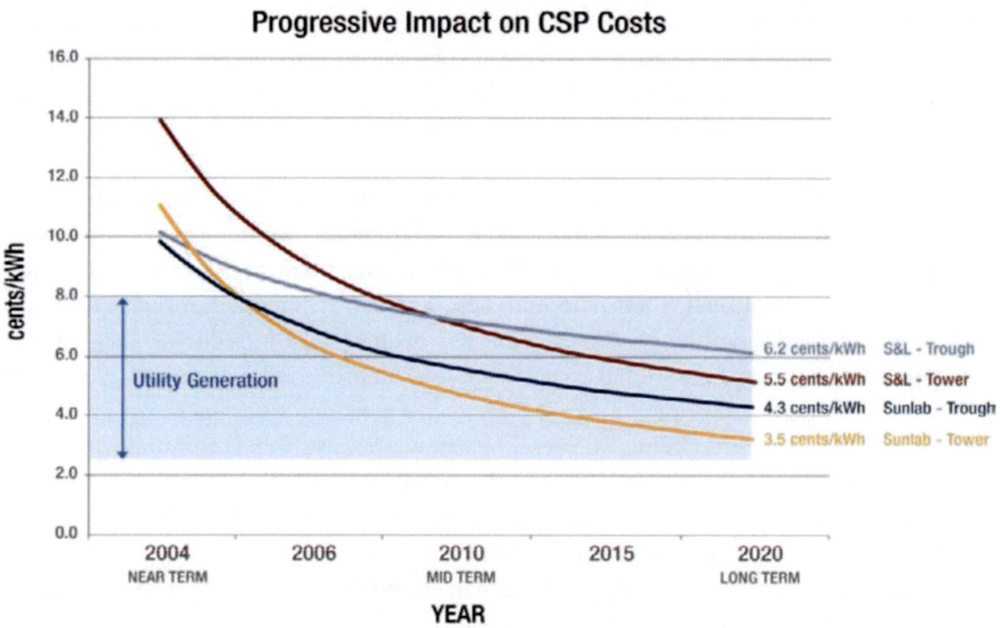

Figure 2-3. Sargent & Lundy CSP Cost Reductions Projections[23]

SAI's Market-Driven Approach

The goals of the SAI are directly market-oriented, in terms of cost, so all R&D activities initiated under SAI will be guided by a systems-based approach. Because the SAI is structured to have an impact on markets in the near-term, the approach will emphasize industry leadership of R&D projects to assure that development planning is driven by commercial business strategies and requirements. Under the SAI, industry-led Technology Pathway Partnerships (TPPs) with universities, component suppliers, and National Laboratories will be established to conduct R&D on all PV system components with a common goal of producing fully-integrated PV systems optimized for U.S. markets.

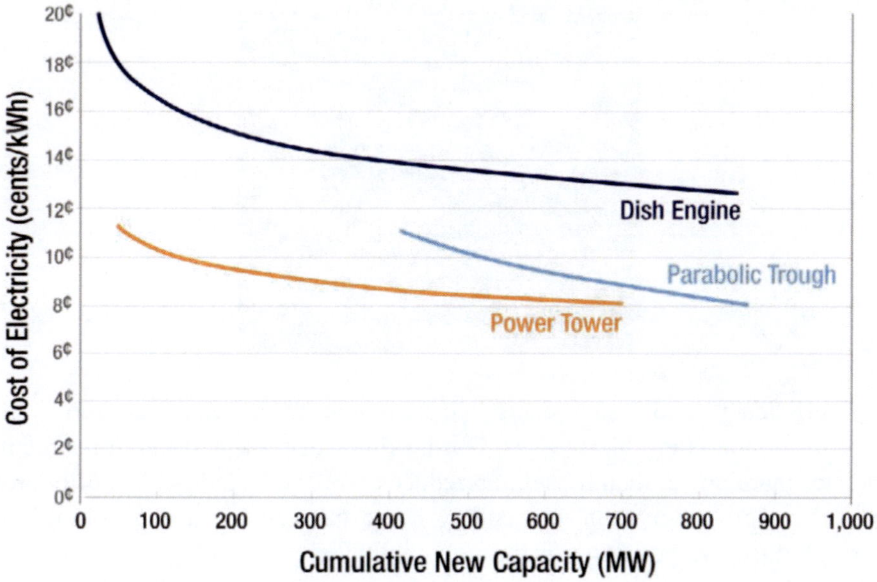

Figure 2-4. Platts/NREL Estimated CSP cost reductions[24]

3.0 Technology Overview

Photovoltaics

Photovoltaic-based solar cells convert sunlight directly into electricity. They are made of semiconducting materials similar to those used in computer chips. When sunlight is absorbed by these materials, the solar energy knocks electrons loose from their atoms, allowing the electrons to flow through the material to produce electricity. The process of converting light to electricity is called the *photovoltaic effect.*

Conventional crystalline silicon PV cells are typically combined into modules that hold about 40 cells; about 10 of these modules are mounted in PV arrays, or panels, that can measure up to several meters on a side. These *flat-plate* PV arrays can be mounted at a fixed angle facing south, or they can be mounted on a tracking device that follows the sun, allowing them to capture the most sunlight over the course of a day. About 10 to 20 PV arrays can provide enough power for a household; for large electric utility or industrial applications, hundreds of arrays can be interconnected to form a single, large PV system. Sunlight can also be focused onto PV panels using lenses or mirrors; such systems are referred to as Concentrating Photovoltaic (CPV), and are far more efficient than their regular PV counterparts.

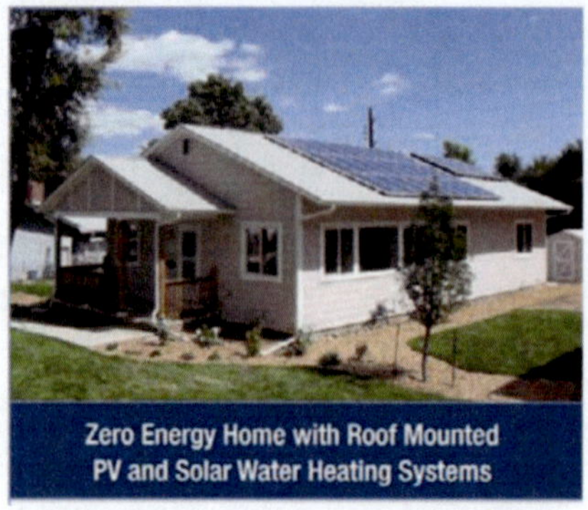

Energy efficiency and solar energy technologies can result in zero net energy consumption from nonrenewable sources. During times of peak demand. a near zero energy home generates mare power than it uses and reduces power demand on the utility provider. In a Florida study, a prototype near zero energy home outperforms a conventional model by providing most of its own power needs throughout the year.

Thin film solar cells use layers of semiconductor materials only a few micrometers thick. Thin film technology has made it possible for solar cells to now serve in building-integrated applications such as rooftop shingles, roof tiles, building facades, or the glazing for skylights or atria. The solar cell version of items such as shingles offer the same protection and durability as ordinary asphalt shingles.

Photovoltaic panels produce direct-current (DC) electricity. Although a number of applications use the direct current from the modules, the fastest-growing markets for PV integrate the panels into systems with power-conditioning equipment that converts the DC electricity to alternating current (AC) electricity. These systems are then interconnected to the utility grid and are referred to as grid-tied systems. Grid-tied and simple DC PV system configurations are illustrated in Figure 3-1 below.

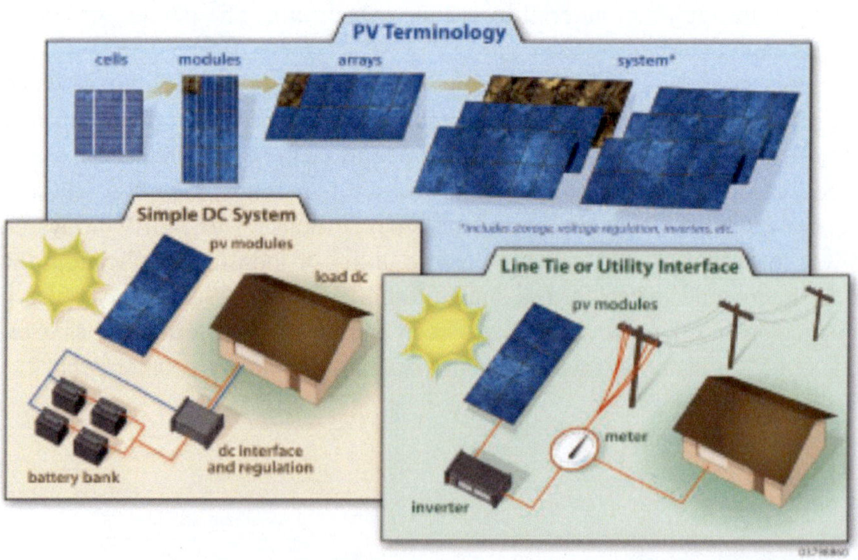

Figure 3-1. Graphical Depiction of PV Systems with Common Terms Illustrated-Source NREL

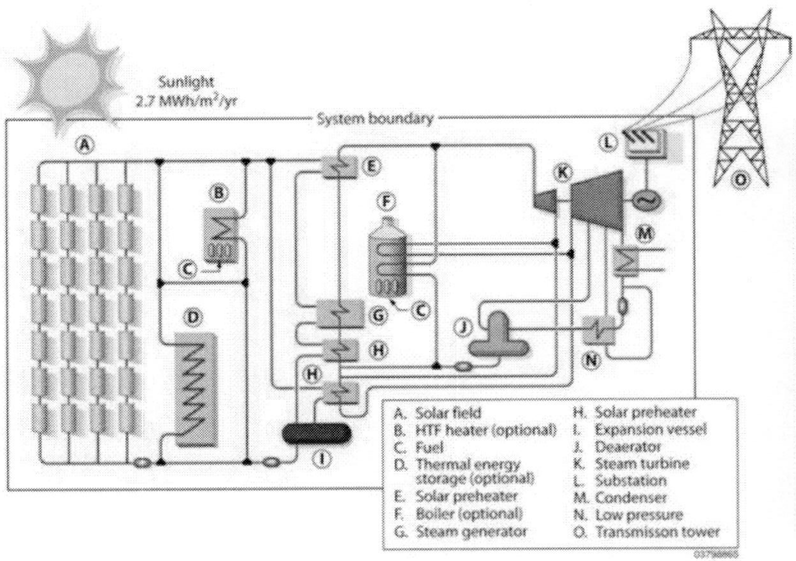

Figure 3-2. Graphical Depiction of CSP Trough (left); Graphical Depiction of CSP Dish-Engine Right)

Source: NREL

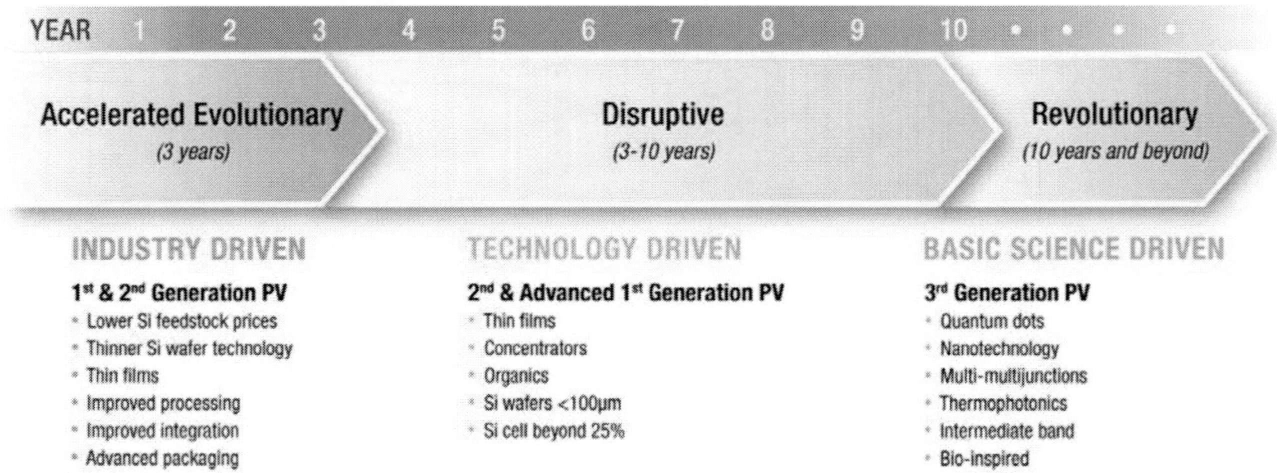

Figure 4-1. Multi-tiered Approach of SAI

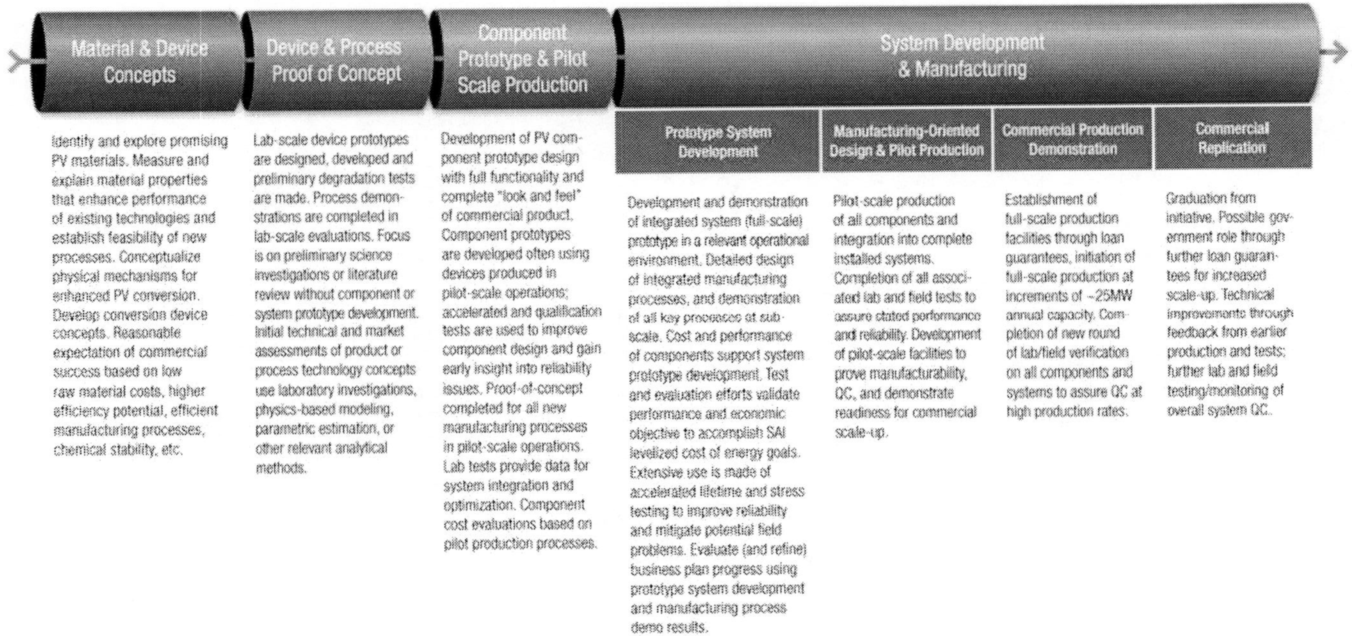

Figure 4-2. Solar Technology Development Pipeline

PHASES	Material & Device Concepts	Device & Process Proof of Concept	Component Prototype & Pilot Scale Production	System Development & Manufacturing	
SOLICITATION	Solar Energy Utilization	Future Generation PV Devices & Processes	PV Component / System Incubator	University Product & Process Development Support	Technology Pathway Partnerships
FUNDING SOURCE	DOE/OS, BES	DOE / SETP	DOE / SETP	DOE SETP	DOE / SETP
DESCRIPTION	New materials and pathways for solar to electric conversion	Novel devices or processes with potentially significant performance or cost advantages	Prototype PV components or systems produced at pilot-scale with demonstrated cost, reliability, or performance advantages	Universities perform targeted materials science and process engineering research in support of industry-led teams developing new PV systems for commercialization in 2010-2015.	PV systems and components ready for mass production delivering energy at target costs
PROJECT LIFECYCLE	3 years	2 years	1.5 years w 9 mo. On/Off Ramp	3 Years	3 years
ANNUAL FUNDING LEVEL	$0.3 - 1.5 M	$0.3 - 1 M	$1 - 2M	Up to $300K/YR	$2 - 7M
TEAM LEADS	Universities or Laboratories*	Small Businesses, Universities, or Laboratories*	U.S. Commercial Entity	Universities	U.S. Commercial Entity
ELIGIBLE PARTICIPANTS	All	All	Universities / Laboratories*	Universities	Universities / Laboratories
ENTRANCE CRITERION	Basic science properties conceived/simulated	Materials synthesized; properties observed	Coupon-scale PV cell; process demonstrated in lab; proof of concept demo	Identification of manufacturing process or component improvements possible through targeted research investigations.	Prototype components; pilot production demo; business case established
EXIT CRITERION	Materials synthesized; properties observed	Coupon-scale PV cell; process demonstrated in lab; proof of concept demo	Prototype components; pilot production demo; business case established	Incorporation of research results into commercial manufacturing operations or product designs.	Commercial PV systems and subsystems; scaled production demonstrated >25MW
TOPICS	• Single-crystal, polycrystalline, amorphous, and nanostructured inorganic and organic materials • Electronic structure • Single or multiple junction solar cells	• New devices and structures using materials such as thin-film silicon, microcrystalline/amorphous silicon, polycrystalline metal chalcogenides and oxides, nanocrystalline materials, biomimetic concepts organic materials, photoelectrochemical cells, dye-sensitized materials, materials with low-dimensional quantum structures • Very-high efficiency epitaxial solar cells or other concepts	• Modules: multiple technologies including CPV) seeking efficient material use, better performance, or improved manufacturing • BOS Components: higher reliability inverters, CPV trackers, rapid installation features storage systems • Systems: controls and smart monitoring, integration of components, factory diagnostics	Identifying and developing: • Fabrication processes to improve material properties during manufacture • Improved solar cell materials • Innovative device designs to improve solar cell efficiency • Simpler lower cost manufacturing processes • New electrical contacting techniques for improved efficiency and reliability • Diagnostic techniques to identify properties and quality of solar cells materials during manufacturing • Improved materials utilization processes • Understanding of chemistry between encapsulants and solar cell materials • Providing careful long-term field testing of modules and systems in support of product improvement	• Partnerships with U.S. industry for projects that focus on development, testing, demonstration, validation, and interconnection of new PV components, systems, and manufacturing equipment • Technology improvements in PV system and component design, integration, and installation will be a focus • Cost reductions, performance enhancements, and reliability improvements are sought for all aspects of PV systems

Figure 4-3. Technology Development Opportunities

CSP Technology

Concentrating Solar Power (CSP) technology does not rely on the photovoltaic effect to create electricity. Instead, CSP systems use mirrors to convert the sun's energy into heat, which in turn is then converted into electricity. Figure 3-2 gives a graphical depiction of *parabolic trough* and *parabolic dish-engine* CSP systems, which are described below.

Parabolic trough systems are the most common form of CSP. They use single-axis-tracking, parabolic trough-shaped reflectors to concentrate sunlight onto an absorber pipe located at the focal line of the parabolic surface. A high temperature heat transfer fluid flowing through the absorber absorbs the thermal energy from the sunlight. Heat in the fluid is then used to make steam in a steam generator. The steam drives a conventional steam-Rankine power cycle to generate electricity. A collector field contains many parallel rows of troughs connected in series. Rows are typically placed on a north-south axis, allowing the single-axis troughs to track the sun from east to west during the day. Trough systems are best suited for large-scale plants, with their extensive arrays of mirrors and large turbine generators.

Parabolic dish-engine CSP systems comprise a solar concentrator and the power conversion unit (PCU). The concentrator consists of mirror facets which form a parabolic dish which redirects sunlight to a receiver mounted on a boom at the dish's focal point. The system uses a two-axis tracker so the concentrator continuously points at the sun. The PCU includes the thermal receiver and the engine-generator. In the thermal receiver, radiant solar energy is converted to heat in a closed hydrogen loop. The heated hydrogen drives the Stirling engine-generator. Because PCUs are air cooled, cooling water is not required, as it is for the large, central power blocks associated with CSP parabolic trough technology. A parabolic dish-engine system using an efficient Stirling engine is shown in the figure above. The modular nature of dish systems allows for them to be deployed in either distributed or remote generation applications, and in large arrays of hundreds or thousands of dishes to produce power on a utility scale.

4.0 Technology Development Strategy

Accomplishing the goals of the SAI will require a multi-tiered, multi-phased program that addresses the near-, mid-, and long-term technological and scientific advances for improved performance, lower cost, and improved reliability of PV system components and installed systems. Figure 4-1 below illustrates, at the highest level, the multi-tiered approach of SAI.

Figure 4-2 shows the solar technology development pipeline and the research and development activities associated with the various stages in the pipeline process.

In the development of advanced PV products, the Solar Program provides numerous technology development opportunities. These opportunities range from research on basic science properties of new materials and processes to the development of new, fully-integrated energy systems ready for mass production. These opportunities are depicted in Figure 4-3 below; specific information about the solicitations can be found on the Solar Program Web site: http://www.eere.energy.gov/solar/

The Office of Science and the Office of Basic Energy Sciences, through the "Solar Energy Utilization" Solicitation, focus on novel technologies that are far from commercialization, such as nanostructured inorganic and organic materials, and multijunction cells. These new and novel materials and pathways for solar to electric conversion are identified, synthesized, and observed. Technologies targeted by this solicitation include single-crystal, polycrystalline, amorphous, and nano-structured inorganic and organic materials as well as single or multiple junction solar cells.

The Solar Energy Technologies Program takes over when a technology is ready for prototype system development/proof of concept. The "Future Generation PV Devices & Processes" solicitation focuses on slightly more mature technologies. These include thin-film silicon, nanocrystalline materials, biomimetic concepts, organic materials, photo-electrochemical cells, dye-sensitized materials, and very-high efficiency epitaxial solar cells, among other concepts.

The "PV Component/System Incubator" solicitation will involve small businesses and non- university research institutes. This solicitation is designed for technologies/processes that have successfully demonstrated a proof of concept/process in a lab, but are not yet mature enough for large-scale commercial production, and the emphasis will be on the barriers to entry towards 2010 commercialization. Prototypes of these PV systems and components will be produced on a pilot-scale in a relevant operational environment with their demonstrated cost, reliability, or performance advantages. PV and CSP modules, components, and systems are all targeted in this phase of the technology development pipeline with the goals that include the more efficient use of materials, better performance, higher reliability, storage systems, and improved manufacturing.

The "University Product and Process Development Support" focuses on targeted materials science and process engineering research in support of industry-led teams developing new PV systems for commercialization in 2010-2015. This research will be conducted by universities.

The last stage of the technology pipeline before commercial replication and graduation from the initiative is targeted by the DOE/SETP's "Technology Pathway Partnerships" (TPPs). These partnerships with US Industry, national laboratories, and universities focus on the development, testing, and demonstration of new PV components, systems, and manufacturing equipment ready for mass production delivering energy at targeted costs.

Industry-Led Technology Pathway Partnerships

The TPP element of the program will establish industry-led teams and provide them the opportunity to apply entrepreneurial creativity and the best industry practices towards the development of innovative technology solutions.

The high-level concept behind the TPP approach is to fund industry-led teams that are comprised of industry members that address the R&D and engineering requirements of much or all of the PV system's value chain.

- The industry-led Technology Pathway Partnerships (TPPs) focus on the near-term, "accelerated evolution" aspects of their technologies, as well as on some of the technical improvements needed in the mid-term for "disruptive" advances toward meeting the 2015 goals. Underlying both the near- and mid-term research is improved scientific understanding of materials, devices, and processes, which can also result in "revolutionary" advances in the longer term (post 2015).

- The driving philosophy of SAI project management will be to ensure maximum value for the DOE/taxpayer investment in solar energy technologies by requiring TPPs to pursue aggressive but realistic technical and production goals, and then holding them accountable to reaching those goals as a condition of continued funding.
- Another key aspect to the TPP approach is the inclusion of two classes of participation. Larger companies/technologies can fully capitalize on their market position and drive those technologies to competitiveness more quickly, while providing emerging companies/technologies the opportunity to vastly accelerate their ability to scale up production and impact the market.

NREL Core Program

The SAI's NREL Core Program will complement the aforementioned TPPs, and is critical to meeting the 2015 goals of the Initiative and facilitating post-2015 technological and global market leadership for the Nation. In the near-and mid-term, the National laboratories and universities will support the TPPs with specific research tasks identified by the individual companies selected. In addition, the NREL Core Program will conduct cross-cutting research that will benefit several companies and technologies by working on fundamental materials, device, and processing issues. Providing measurements and characterization support to industry is one important aspect of this support. Another key element is to support research that keeps the pipeline of innovation full in order to achieve the progress needed to accomplish the 2015 goals, as well as facilitate post-2015 global competitiveness for U.S. PV technologies. Traditionally, this research has often been performed by researchers at National laboratories and universities, as well as by new corporate start-ups (small and large businesses) that are a few years away from qualifying as a TPP participant, but have excellent prospects for meeting the longer-term and post-2015 goals.

The teamed collaboration and technical integration that results among in-house researchers and university and industry subcontractors has been the hallmark of successful programs such as NREL's Thin Film PV Partnership. This has resulted in today's U.S. global leadership position for thin-film technologies in laboratory and the marketplace.

As part of its Core Program, NREL has the Electronic Materials and Devices (EM&D) project that performs research in semiconductor, dielectric, conductor, encapsulant materials, and device properties. For Silicon Materials and Devices, NREL will develop and transfer technology to enable processing of high-performance crystalline and thin-film devices. In regards to Thin-Film Polycrystalline Compounds, NREL seeks to Adapt high efficiency CIGS (Copper Indium Gallium Diselenide) and CdTe (Cadmium Telluride) solar cell materials systems to implementation in low-cost manufacturing and durable performance. NREL will also work to develop a suite of processes and processing tools that operate rapidly at atmospheric pressure and low temperature. By 2015, NREL will have also developed and transferred technology for production of 40%-efficient concentrator cells that are manufacturable at competitive costs. Work will also continue involving cell and module stability and reliability, with a goal of elucidating the factors that contribute to cell and module degradation in order to develop models that predict service life at the prototype cell and module stage by 2015. In collaboration with the DOE Office of Science, the program will also conduct basic research to target breakthroughs in key areas such as ultra-high efficiency and/or ultra-low cost materials and devices, and solar- hydrogen conversion and storage technologies.[25]

Science and Technology Facility (S&TF)
located at the National Renewable Energy Laboratory in Golden, CO

The centerpiece of the building is the Process Development and Integration Laboratory (PDIL), specifically designed to accommodate a new class of crystalline silicon and thin-film PV processing and characterization tools. The basic concept of the PDIL is to allow researchers to pass samples between equipment and processing steps in a controlled way, avoiding contamination from the air. The PDIL will also allow a scientist to integrate control systems and databases in such a way that someone who is growing a sample can see the results of a measurement and vice versa. The S&TF will include nine advanced materials synthesis, characterization, and general support laboratories.

Coordination with the Office of Basic Energy Sciences

The solar energy utilization research supported by the Office of Basic Energy Sciences is focused on advancing fundamental understanding of atomic/molecular level interactions and reactions associated with the conversion of solar energy into electricity and chemically stored energy. To ensure the success of SAT, the program is developing linkages in order to leverage these activities through collaboration and coordination – for example, workshops, planning, and joint support actions. Starting in FY07, new research activities will be initiated through recent solicitations for awards in the following areas:

For *solar to electric energy conversion,* the major emphases will be on the synthesis and discovery of new materials to efficiently absorb sunlight and new techniques to harness the full spectrum of wavelengths in solar radiation. The major activities include:

- Recent advances in the design and synthesis of nanostructured architectures offer great promise to revolutionize the technology used to produce solar electricity. Major solar electricity research activities will be initiated in the following areas:
    - Exploiting thin films, organic semiconductors, dye sensitization, and quantum dots for cheaper, more efficient, and longer-lasting solar PV cells.
    - Synthesis and fabrication of nanoscale architectures by novel top-down and bottom-up techniques.

- Nanoscale characterization using electron, neutron, and X-ray scattering and spectroscopy with an emphasis on ultrafast tools.
- Simulations of electronic and molecular behavior in nanoscale semiconductor assemblies using density functional theory.

- These activities are aimed at paving the foundation to drive a revolution in the way that solar cells are conceived, designed, implemented, and manufactured.

For *solar fuels research,* the major activities include:

- Application of the revolutionary advances in biology and biotechnology to the design of plants and organisms that are more efficient energy conversion "machines," and
- Design of highly efficient, all-artificial, molecular-level energy conversion machines exploiting the principles of natural photosynthesis.

A key element in both approaches is the continued elucidation—by means of structural biology, genome sequencing, and proteomics—of the structure and dynamics involved in the biological conversion of solar radiation to sugars and carbohydrates. The disclosure of these long-held secrets of natural solar conversion by means of cutting-edge experimentation and theory will enable a host of exciting new approaches to direct solar fuel production.

Artificial nanoscale assemblies of new organic and inorganic materials and morphologies, replacing natural plants or algae, will be synthesized to use sunlight to directly produce H_2 by splitting water and hydrocarbons via reduction of atmospheric CO_2. The main emphasis of this activity is to develop approaches and systems to bridge the vast knowledge gap in generating durable, cheap, and efficient artificial molecular machines to convert sunlight to chemically stored energy.

In solar thermal systems, the main research activities focus on novel thermal storage materials with an embedded phase transition to offer the potential of high thermal storage capacity and long release times, bridging the diurnal cycle. Nanostructured thermoelectric materials, in the form of nanowires or quantum dot arrays, will also be investigated for direct electricity production from temperature differentials with efficiencies of 20-30% over a temperature differential of a few hundred degrees Celsius. The much larger differentials in solar thermal reactors make even higher efficiencies possible. New low-cost, high-performance reflective materials for the focusing systems are needed to optimize the cost effectiveness and continued improvement of all concentrated solar thermal technologies.

5.0 Program Activities

Systems Analysis

The Systems Analysis element of the DOE's Solar Energy Technologies Program will devote its resources to evaluating all program activities to ensure a well-managed, market-based technology research, development and deployment organization. The establishment and tracking of milestones will serve the Solar Program as it works to achieve its goals.

The Solar Program will partner with the U.S. PV and CSP industries to improve and advance the deployment of systems design tools, component and systems performance databases, systems performance prediction tools, and market penetration models. The Systems Analysis activities will take full advantage of relationships and access to information offered by the monitoring of selected SAI Technical Pathway Partnerships, and other solicitations.

Defense Advance Research Project Agency (DARPA) Advances Solar Technologies Outside of DOE, the Defense Advanced Research Projects Agency (DARPA) within the Department of Defense recently began an ambitious project to develop a 50% efficient solar electric device in four years for use in recharging batteries for critical information technology systems under battlefield conditions.

In February 2005, the Defense Advance Research Projects Agency's (DARPA) Advanced Technology Office issued a solicitation for up to $53 million for the Very High Efficiency Solar Cell (VHESC) program.

The objective of this program is to demonstrate at least 50% efficiency in a PV device. The program requires that teams be formed to comprehensively address all aspects of the high-efficiency PV problem including the development of fabrication processes that are scaleable to industrial manufacturing and an affordable product.

A broad consortium led by the University of Delaware—including 15 universities, corporations, and laboratories—was selected.

DOE is collaborating with DARPA in its selection of contractors and a technology portfolio, its creation of workshops, exchanges, and requirements for systems analysis. NREL is a DARPA contractor on this program.

The Solar Program has developed a set of Technical Improvement Opportunities (TIOs) that generate a framework for the TPPs to identify the technical requirements that they need to pursue in order to meet program objectives. The TIO framework is depicted in Figure 5-1 below with Tier 1 TIOs consisting of Modules; Inverters and Balance of Systems (BOS); Systems Engineering and Integration; and Deployment Facilitation.

The Solar Program will continue to use the Levelized Cost of Electricity (LCOE) as the SAI's primary metric. The LCOE incorporates all of the critical technical elements of a PV system related to performance and reliability, while also capturing manufacturing costs. The LCOE is an elegant metric in that it allows applicants to pursue a variety of innovative combinations of improved performance, reliability, and cost that will reach LCOE targets. Many of the technology pathways and improvements to realize SAI LCOE targets have been addressed or demonstrated at the laboratory level. The challenge of the SAI is to more quickly move laboratory findings into commercially available products.

The Systems Analysis activities will include two main elements:

- Program Evaluation and Stage Gate Reviews
- Benchmarking & Design Support Tools

TIOs		Metrics			
TIER 1 TIOs	TIER 2 TIOs	Performance Efficiency	Cost	O&M	Reliability
Modules	Module				
	Absorber				
	Cells and Contacts				
	Interconnects				
	Packaging				
	Manufacturing				
Inverter & BOS	Inverter				
	Inverter Software				
	Inverter Components/Design				
	Inverter Packaging/Manufacturing				
	Inverter Integration				
	Other BOS				
Systems Engineering & Integration	System Engr. & Integration				
	System Manufacturing/Assembly				
	Installation & Maintenance				
Deployment Facilitation					

03798834

Figure 5-1. Technology Improvement Opportunities. PV Color boxes indicate degree of impact that TIO has on each metric and overall levelized cost of energy. (red-high, yellow-medium, white-low)

Program Evaluation & Stage Gate Reviews

The Stage Gate process (see Figure 5-2) is a means of making focused, goal-oriented disciplined research and development decisions at all stages along a product development cycle. The Stage Gate process is being used throughout the Solar Program.

The basic approach is as follows: project development takes place in a series of sequential "stages" and the reviews of the projects are done at "gates." The "development cycle" for a project can go from applied research investigations through technical activities that support massive production.[26] Increasing rigor, industry involvement, and expenditure are expected as development moves through successive stages. This approach is applied across varying levels of development and integration of new solar systems, from new subcomponents to fully integrated systems.

Implementation of the Stage Gate process is allowing the Solar Program to achieve important improvements in several principal areas of R&D portfolio management:

- Strategic Alignment: R&D activities are directly relevant to the overall goals and strategy of the solar program
- High Value: R&D portfolio positioned to make the largest potential impact on the solar energy industry

- Balance: Technical activities can be expected to produce cost, reliability, and performance improvements in all important areas of solar systems development.
- Accountability and Transparency: allows program to establish accountability for progress in meeting quantified, integrated goals, which can be readily seen and assessed by stakeholders

Current and future market considerations, quantification of the impact of technical improvement opportunities, and prioritization of program activities are combined in a dynamic decision- making process wherein the list of active projects is continually revised.

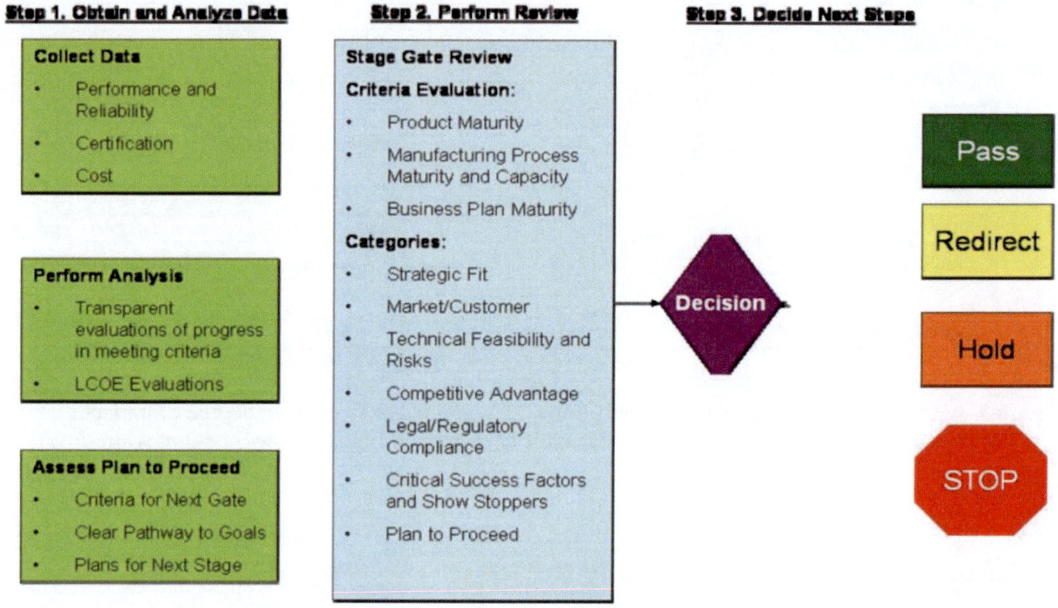

Figure 5-2. Stage Gate Review Process

Benchmarking and Design Support Tools

Benchmarking capabilities and design support tools are key to program, and ultimately, industry success. The program must track its progress towards key targets, especially the SAI Stretch Targets. Benchmarking component and system cost, performance, and reliability and using these data in models is necessary to verify progress. The ongoing analysis of technology, market, and cost trends will help us understand and continually incorporate improvements that can be expected to occur by 2015 as a result of program activities. Industry also requires these capabilities because as solar industry technologies mature in the market, customer expectations for performance, cost, and reliability of systems are increasing rapidly.

- The primary objective of the benchmarking activity is to document historical and current cost, reliability and performance metrics for fielded systems and components.

- Benchmarked data will be used to support the model development and analysis, to help evaluate project proposals and progress under the SAI solicitations, and to assist in Solar program planning (input to multi-year plans, etc.).

In PV, for example, there is a strong trend away from capacity-based specifications to performance-based guarantees over extended periods. The ability to readily and accurately predict a system's performance and LCOE is becoming critical in ensuring a company's viability/profitability. Additionally, a shift away from component-based program planning to more of a systems perspective demands a comprehensive set of predictive modeling tools that allow for a variety of parametric/sensitivity studies. Systems perspective helps to sort out component development needs, priorities, and informs R&D decision making.

SDA Modeling and Analysis

Over the last three years the Solar Program has developed and employed a Systems-Driven approach (SDA) to program management, which uses empirical data to assess the relative marketplace impacts of different research and development directions. This approach applies throughout the solar system value chain, from fundamental research on materials to improved components and integrated solar systems. The Stage Gate methodology, the modeling and analysis work, and the collection of sound empirical data through testing and evaluation, are interconnected efforts employed as part of the Systems-Driven Approach.

SAM Model

A DOE/NREL/Sandia team has developed the Solar Advisor Model (SAM) to provide a common platform for calculating LCOE and to conduct TIO sensitivity studies on overall system LCOE. The SAM Model is currently undergoing validation, user-testing and refinement. The model allows analysts to investigate the impact of variations in performance, cost, and financial parameters on key figures of merit. The model is intended for use by DOE, laboratory management, and research staff for program planning purposes. The model may also be used by members of the solar industry to inform their internal R&D direction and to estimate systems cost and performance.

Photovoltaic Systems

Increasing PV system efficiency and lowering manufacturing costs are essential for the success of the Solar America Initiative. Program activities will focus on bringing emerging technologies and manufacturing processes out of the laboratory and to market. Ongoing and planned R&D includes the following activities:

PV Cell Technologies

- Develop and demonstrate fabrication of high-efficiency crystalline silicon, thin film, III- V multijunction, and future-generation cells. The specific efficiency milestones for the various cell technologies are reflected in the Gantt chart.
- Develop novel c-Si cell-contacting schemes.

- Engage in efforts to transfer laboratory cell performance to production modules.
- Assess and verify durability and reliability of future-generation solar cells.

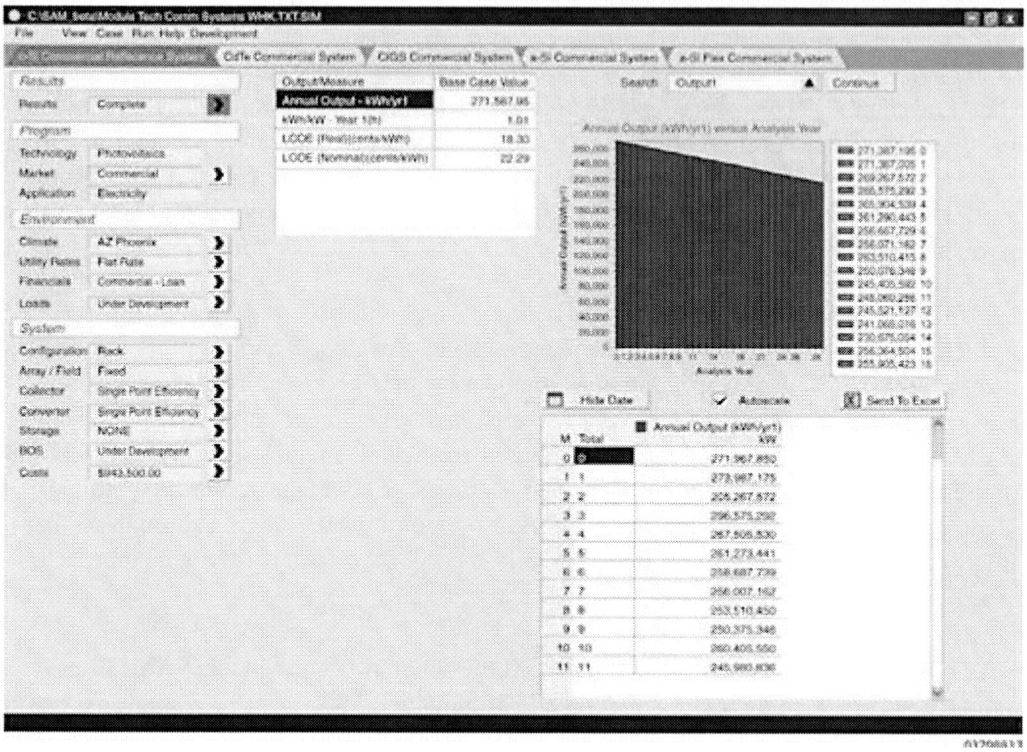

Figure 5-3. SAM Model Screen Shot

PV Module Technologies

- Pursue innovations to improve manufacturability of c-Si cells and interconnects.
- Develop and fabricate amorphous silicon, CIGS, and CdTe modules that have increased solar-to-electric efficiencies.
- Pursue innovations in thin film module design leading to reduced active-area losses through edge and interconnect scribes.

PV Manufacturing R&D Program

- Work with industry partners on improved manufacturing processes.
- Implement fully the Science and Technology Facility (S&TF) tool suites to facilitate laboratory/industry interaction in developing manufacturing technologies and accelerating technology transfer to industry.

- Identify commercialization pathways for promising new technologies via university/industrial partnerships.

PV Inverter Technologies

- Work with industry partners via Technology Pathway Partnerships, and specifically with GE and Xantrex through the High Reliability Inverter Initiative.

PV Systems Development

In the early stages of solar energy R&D, the Solar Program has necessarily been focused on system components (primarily PV cells and modules), but under the SAI, the focus will be on solar energy systems development. This will ensure the most direct path to the marketplace for system and technology advances. Under this portion of the SAI, TPPs will be focused initially on photovoltaic systems and the PV industry. It is anticipated, however, that teams will also include university partners and other laboratory expertise that capitalizes on the Solar Program's historic investment in the U.S. solar energy R&D infrastructure.

To integrate the program into the current industry structure and status, the following program phasing will be pursued:

Phase 1: Demonstrate Pilot Production of Lower Cost Systems

- R&D Project Duration of 3 Years (2007-2009)
- Two classes (Systems and Subsystems) give companies the flexibility to align projects with their targeted business model, technical capabilities, and product/process maturity
- Systems Class: Development of Turnkey Systems
    - DOE Support for Awards limited to $20 million over the three-year period ($10 million/yr ceiling for DOE Award Support + 50% minimum Cost Share; 4-10 Awards expected)
  - Subsystems Class: System-driven R&D on Components
    - DOE Support for Awards limited to $8 million over the three-year period (Up to $4 million/yr ceiling for DOE Award Support + minimum 50% Cost Share, 10-15 Awards expected)
- Down-select following intermediate Stage Gate (After 16-18 Months)

Phase 2: Demonstrate Replication of Low Cost Pilot Production

- R&D Project Duration of 3 Years (2010-2012), with new entrants eligible.
- Classes and requirements TBD based on Phase 1 program experiences.
- Possible use of different procurement instruments (e.g. Loan Guarantees), depending on maturity of technology and business models

Phase 3: Develop Supply Chain & Distribution Infrastructure

- Reduce capital expenditure intensity of manufacturing scale-up, through targeted investments in equipment/feedstock suppliers, distribution channels, etc.

Concentrating Solar Power Systems

In parallel to the PV R&D efforts, the Solar Program will continue to pursue advances in driving CSP to be cost competitive. Projections for CSP cost trends are shown in Figure 5-4.

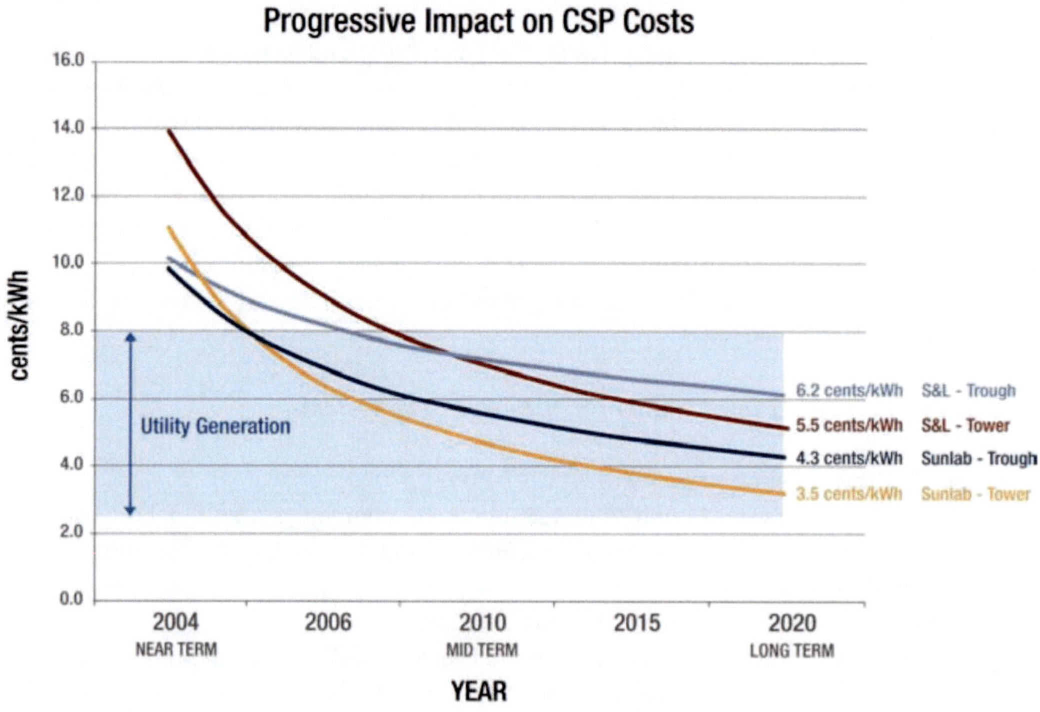

Figure 5-4. Projections for LCOE Cost Reductions for CSP by Sergeant and Lundy and DOE's SunLab[27]

Until the CSP budget enables additional contracts that augment core R&D lab activities, the Solar Program will focus on the following opportunities:

Troughs

- Provide technical support for CSP projects in the Southwest, including optical testing to optimize receiver and concentrator designs
- Begin development of next-generation system capable of operating at 450° Celsius
- Develop improved receiver testing capabilities, advanced selective coatings, receiver maintenance systems, and optical characterization tools

Dish/Engines

- Work with industry on the six-dish mini-plant at the industry/laboratory test facility in New Mexico to test dishes in a power plant configuration
- Focus on engineering solutions to Stirling engine reliability issues
- Work with industry to improve dish manufacturability and any necessary component upgrades or redesigns
- Assist industry in the design of the 1.0 MW demonstration project in California and subsequent projects

Storage

- Identify a heat transfer fluid suitable for both the solar field and storage system at temperatures up to 450°C; the fluid must have a high thermal capacity, low vapor pressure, and remains liquid at ambient temperatures.
- Conduct research, lab tests and field demonstrations of thermal energy storage systems that may offer a near-term, low-cost storage option for industrial trough projects

Technology Evaluation

The Technology evaluation (TE) activities within the Solar Program are focused upon collaboration and cooperation with industry and on obtaining data that quantify performance improvements within the industry and the Solar Program. The PV TE staffs currently obtain "hardware" from a variety of sources (U.S. and global manufacturers, system integrators, and university/internal researchers) and will obtain both prototype and commercial hardware from TPP teams. Testing equipment is very expensive and can be cost prohibitive, but through the SAI's TE activities, DOE is ensuring that companies do not have to make investments in such equipment and can instead focus their resources elsewhere.

- TE activities will include several different types of tests on systems, components, or devices. These evaluations produce performance parameters, validated data base entries, data reports and analyses, test protocols, and data analysis methodologies.
- In addition, in-line diagnostics for use in product improvement or process understanding are developed.
- Specialty measurements can be performed to support evaluation of new system or component concepts; often the insights gained in these evaluations lead to improvements in test protocols.
- Access is provided for direct industry use of "one of a kind" test facilities, such as the new Science and Technology facility at NREL or the PV Systems Optimization Lab at Sandia.
- All results are used in direct support of component and manufacturing process improvement, integration of systems, stage gate evaluations, benchmarking, codes

and standards development, and performance modeling. Results are also valuable in planning and evaluating Solar Program technical activities.

Supporting these Solar Program technology evaluation efforts are the personnel, processes, capabilities, and facilities at several National Laboratories (National Center for Photovoltaics: National Renewable Energy Laboratory and Sandia National Laboratories) and the Solar Energy Technology Program's Regional Experiment Stations (SW, Southwest Technology Development Institute and SE, Florida Solar Energy Center). This work is complementary to and supportive of the capabilities of many private entities and standards and codes organizations that also support the PV industry. TE support for CSP is centered at the National Solar Thermal Test Facility located in Albuquerque, NM at the Sandia National Laboratories.[28]Technology evaluation results are fully documented for integrity and transparency, with a formal protocol for product delivery, testing, and dissemination of results. Proprietary data are routinely handled without compromise.

Market Transformation

Market Transformation is the set of SAI activities that aims to reduce market barriers and promote deployment assistance for solar energy technologies through non-R&D activities. Funded through separate solicitations in parallel to the Technical Pathway Partnerships, these activities will augment the R&D efforts of industry, National Laboratories, and universities.

The SAI goal of achieving cost-competitiveness of solar energy technologies by 2015 across all market sectors is an aggressive one, and will require both technologically-based solutions and consideration of non-R&D barriers to full-scale market penetration. As the previous chapters have discussed, there are many pathways for improving solar energy technologies using R&D, manufacturing engineering or other technical solutions. These technologically-based approaches offer opportunities for substantial progress in reducing the cost of solar technologies.

Yet the picture is incomplete without consideration of the non-R&D or institutional barriers that prevent full-scale market penetration of solar technologies. Inability to easily connect to the electric grid, lack of net-metering regulations, or the forced-purchasing of liability insurance by a utility, can all prevent the installation of a PV system, even one that is highly-efficient at a low- cost. In that case, despite the significant gains realized through R&D, our Nation will not accrue any energy, security, nor environmental benefits. It is therefore critical to address infrastructure- related market barriers to solar technologies in addition to R&D work.

Market Transformation Planned Activities

In conclusion, the Market Transformation element will work in parallel to the SAI TPPs to reach a broad spectrum of industry and stakeholders. Market Transformation and Technology Pathway Partnerships will help ensure success for the SAI by covering a wide range of technical and nontechnical barriers.

Figure 5-5 shows the various Funding Opportunity Announcements (FOA's) for the Market Transformation activities of the Solar America Initiative.

MARKET TRANSFORMATION FUNDING OPPORTUNITY ANNOUNCEMENTS (FOA'S)

SOLICITATION	Solar Codes and Standards			State and Utility Solar Technical Outreach		Solar City Strategic Partnerships	Solar America Showcases
TOPIC AREA	Solar Codes and Standards Working Group Leadership	Experience-Based Utility PV Capacity Credits	National Voluntary PV Module Standard	State Solar Technical Outreach	Utility Solar Technical Outreach		
DESCRIPTION	A central body to address, as appropriate, all solar codes and standards issues that are encountered by DOE, DOE contractors, partners, award recipients and stakeholders. The group will assist state and other regulating and inspection bodies in adopting best practices, guiding stakeholders, and in verifying the adequacy of particular equipment and installation methods. It will also work with utilities and state/local jurisdictions to standardize interconnection and installation requirements for grid-connected PV systems.	This Topic Area will support the development and adoption of specific capacity credit values or other utility-approved measures for system benefits of PV. The main deliverable will be a report, paper, software product or other medium that contains the methodology to be used for calculating capacity credit values for PV Systems of widespread use to utilities and other solar stakeholders.	The main focus of this Topic Area is the creation of a national voluntary PV module rating standard, including performance, reliability, safety, anticipated degradations and operational limits and the establishment of testing procedures and protocols for use. The purpose is to increase consumer confidence in PV module performance by bringing actual module performance in line with module nameplate ratings.	Will identify and promote efforts by the relevant decisions makers to accelerate the adoption of solar technologies within their states. Where appropriate, DOE will expect coordination among the various organizations throughout the award period to best address shared solar issues that affect multiple groups of State decision makers.	Will provide targeted solar information, education, and resources to electric utilities. Also will strengthen outreach to utilities who already maintain advanced solar programs and assist such utilities in advancing their acceptance and promotion of solar even further. Aims to foster strong strategic relationships with utilities on solar issues to identify and provide solutions to strategic market barriers to solar energy technology adoption and enable key commercialization activities to promote the SAI.	Support for cities ready to take a comprehensive, systematic, city-wide approach to solar technology that facilitates mainstream adoption and provides a model for other cities. DOE is looking for cities to commit to achieving a sustainable solar infrastructure, not simply a year or two of experimental solar projects. Recipient Cities are expected to develop a city-wide solar implementation plan as a deliverable under this project.	DOE will provide technical assistance to large-scale, high-visibility solar installation projects that have the ability to impact the market for solar technologies through large project size, use of a novel solar technology, and/or use of a novel application for a solar technology. Ideal projects will have a total capacity in excess of 100kw. (Projects may include multiple sites and do not have to be co-located.)
FUNDING CEILING	$4,200,000	$100,000	$3,000,000	$225,000	$900,000	$200,000	$2,700,000 (total FY07 funding)
FUNDING FLOOR	$500,000	$25,000	$500,000	$50,000	$50,000	$80,000	
# OF RECIPIENTS	1-2	1-2	1-2	4-6	1-2	6-10	Up to 25
PROJECT LIFECYCLE	5 years	1 year	3 years	3 years	3 years	24 months	

Figure 5-5. Market Transformation Funding Opportunity Announcements (FOA's)

Program Activities Chart

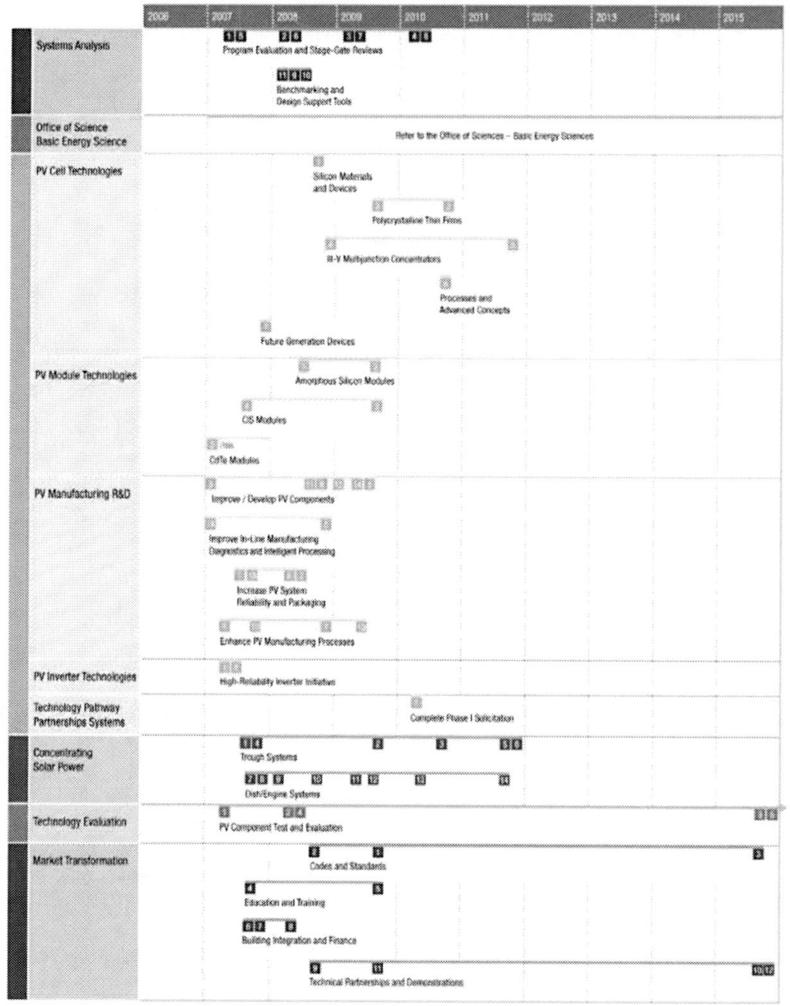

	MILESTONE	DATE
	SYSTEMS ANALYSIS	
1	Validate and characterize Technology Pathway Partnership (TPP) targets and metrics	04/05/2007
2	TPP Stage-Gate review and down-selects	02/05/2008
3	TPP Stage-Gate review and down-selects	02/05/2009
4	TPP Stage-Gate review and down-selects	02/05/2010
5	Validate and characterize NREL EM&D targets and metrics	04/05/2007
6	NREL EM&D Stage-Gate review and down-selects	02/05/2008
7	NREL EM&D Stage-Gate review and down-selects	02/05/2009
8	NREL EM&D Stage-Gate review and down-selects	02/05/2010
9	Develop improved PV systems performance and cost model	04/01/2008
10	Develop public-access PV design support database	04/01/2008
11	Develop open-source PV manufacturing cost model	02/01/2008
	OFFICE OF SCIENCE BASIC ENERGY SCIENCE	
	Refer to the Office of Sciences – Basic Energy Sciences	
	PHOTOVOLTAICS	
	PV Cell Technologies	
1	Develop 20% Silicon SHJ Champion cell	10/01/2008
2	Develop 20% CIGS Cell: 17% CdTe cell	09/03/2007
3	Develop 12% dye-sensitized cell	10/02/2009
4	Develop 40% MJ Concentrator	10/06/2008
5	Demonstrate 33% module efficiency with 42% efficient cell	10/03/2011
6	Demonstrate atmospheric processing for CIGS cells	09/30/2010
7	Demonstrate 1% polymer-based solar cell	10/01/2007
	PV Module Technolgies	
1	Develop > 11% Cell	08/12/2008
2	Fabricate modules at 25 MW capacity at 10%	08/12/2009
3	Demonstrate 25 % increases in CIGS deposition to 15-in-min	08/12/2009
4	Develop low-cost process for TCO coating	08/13/2007
5	*Milestones to be determined*	TBD
	PV Manufacturing R&D	
1	Verify 25% reliability increase for 500-kW inverter	5/30/2007
2	Complete protection-system production line scale-up and calibration	6/13/2008
3	Achieve 30% labor reduction in installation with new materials handling tools	1/31/2007
4	Demonstrate 2% module efficiency increase	6/30/2009
5	Implement in-line ellipsometer for production process monitoring	12/31/2008
6	Demo 30-yr module reliability and 50% increase in energy production/area	5/14/2008
7	Complete demo of manufacturing methods for 170W 11% Silicon-Film module	12/31/2008
8	Demo Cz-Si module manufacturing <$2/W &>15% efficiency	9/30/2008
9	Develop 25 wafers cut with 20% thinner cutting wire and no yield change	3/30/2007
10	Demo 15% efficient cells manufactured on production line	8/31/2007
11	Complete production-ready controller and drive system for CPV tracker	6/30/2008
12	Complete market study and submit PV roof tile module for UL listing	1/30/2009
13	Complete cell line factory design	5/29/2009
14	Complete qualification and testing of high-voltage module	5/29/2009
15	Complete development of Computer Integrated Manufacturing System	1/15/2007
16	Complete long-term exposure testing of modules w/ thin-film specific encapsulant	6/1/2007
	PV Inverter Technologies	
1	Final Report on Inverter Development (GE)	03/30/2007
2	Final Report on Inverter Development (Xantrex)	03/30/2007
	Technology Pathway Partnerships Systems	
	Complete Phase I Activities	02/03/2010

	MILESTONE	DATE
	CONCENTRATING SOLAR POWER	
1	Performance: Demonstrate 78% thermal efficiency in receiver	09/30/2007
2	Performance: Demonstrate 82% thermal efficiency in receiver	09/30/2009
3	Performance: Demonstrate 70% optical efficiency in collector	09/30/2010
4	Cost: Demonstrate trough system cost @ 11-13¢/kWh	9/28/2007
5	Cost: Demonstrate trough system cost @ 10-12¢/kWh	9/30/2011
6	Cost: Demonstrate thermal storage @ $20/kWh	9/30/2011
7	Integrate/validate component models	9/30/2007
8	Complete market requirements evaluation	12/31/2007
9	Upgrade reliability methodology	3/31/2008
10	Start testing of SES Gen 3 System	9/30/2008
11	Integrate/validate new component models	6/30/2009
12	Develop project operations database	9/30/2009
13	Review and update market evaluation	6/30/2010
14	Start testing of advanced system	9/30/2011
	TECHNOLOGY EVALUATION	
1	Develop system performance rating protocol	3/16/2007
2	Demonstrate improved system performance rating protocol	3/17/2008
3	Provide independent testing to TPPs and industry companies	9/30/2016
4	Demonstrate new accelerated test protocols for thin film modules	4/11/2008
5	Provide independent testing to TPPs and industry companies	9/30/2016
	MARKET TRANSFORMATION	
1	Complete IEEE/IEC Voluntary Module Performance Testing Standard	9/30/2009
2	Complete study of Utility Solar Capacity Credit Valuations	9/28/2007
3	Complete removal/minimization of all code/standard barriers to PV market penetration	9/30/2015
4	Complete "Train the Trainer" for at least ten trainers	9/28/2007
5	Complete federal involvement in training/certification for PV	9/30/2009
6	Conduct Solar Decathlon 2007	9/28/2007
7	Complete initial finance study	9/28/2007
8	Release financial "Best Practice" and utility/financial institution Solar Pathway instrument guidance	3/28/2008
9	Complete City Strategic Partnership City Energy Plans from Phase I	9/26/2008
10	Complete Solar America City Program	9/30/2015
11	Provide final technical assistance to Solar America Showcases Phase I	9/26/2008
12	Complete Solar America Showcases Program	9/30/2015

APPENDIX A. ACRONYMS

AC	alternating current
a-Si	amorphous silicon
a-Si:H	hydrogenated amorphous silicon
BIPV	building-integrated photovoltaics
BNL	Brookhaven National Laboratory
BOP	balance of plant
BOS	balance of systems
Btu	British thermal unit
c-Si	crystalline silicon
CdTe	cadmium telluride
CIGS	copper indium gallium diselenide
CIS	copper indium diselenide
COE	cost of energy
COSE	cost of saved energy
CPV	concentrator photovoltaics
CRADA	cooperative research and development agreement
CSP	concentrating solar power
CY	calendar year
DC	direct current
DOD	U.S. Department of Defense
DOE	U.S. Department of Energy
EERE	DOE Office of Energy Efficiency and Renewable Energy
EIA	Energy Information Administration
EPA	Environmental Protection Agency
EPAct	Energy Policy Act of 2005
ES&H	environment, safety, and health
F SEC	Florida Solar Energy Center
FY	fiscal year
GaInNAs	gallium indium nitrogen arsenide
GO	Golden Field Office
GPRA	Government Performance Results Act
GW	gigawatt
GWp	peak gigawatt
HALT	highly accelerated lifetime testing
HIT	heterojunction with intrinsic thin layer
HTF	heat-transfer fluid
IEA	International Energy Agency
IEEE	Institute of Electrical and Electronics Engineers
IPP	independent power producer
kW	kilowatt
kg	kilogram
kWe	kilowatt electric
kWh	kilowatt-hour

kWht	kilowatt-hour thermal
LCOE	levelized cost of energy
LEC	levelized energy cost
LED	light-emitting diode
m2	square meter
MMBtu	million Btu
MOS	measure of success
MPPT	maximum power-point tracking
MTBF	mean time between failure
MTBI	mean time between incident
MYPP	Multi-Year Program Plan
MYTP	Multi-Year Technical Plan
MW	megawatt
MWe	megawatt-electric
NAS	National Academy of Sciences
NASA	National Aeronautics and Space Administration
NCPV	National Center for Photovoltaics
NEC	National Electrical Code
NEMS	National Energy Modeling System
NRC	National Research Council
NREL	National Renewable Energy Laboratory
NSTTF	National Solar Thermal Test Facility
NTRC	National Transportation Research Center
O&M	operations and maintenance
PDIL	Process Development and Integration Laboratory
PV	photovoltaics
PWF	present worth factor
R&D	research and development
RFP	request for proposal
RPS	renewable portfolio standard
S&TF	Science and Technology Facility
SAM	Solar Advisor Model
SBIR	Small Business Innovative Research
SCADA	supervisory control and data acquisition
SDA	systems-driven approach
SEP	State Energy Program
SERES	Southeast Region Experiment Station
Si	silicon
SINC	Systems Integration and Coordination (Team)
SNL	Sandia National Laboratories
SRCC	Solar Rating and Certification Corporation
STTR	Small Business Technology Transfer Research
SWRES	Southwest Region Experiment Station
SWTDI	Southwest Technology Development Institute
TBD	to be determined
TES	thermal energy storage

TIO	technology improvement opportunity
TMY	typical meteorological year
UV	ultraviolet
W	watt
Wp	peak watt
WGA	Western Governors' Association
ZEB	Zero Energy Buildings
ZEH	zero energy home

APPENDIX B. CONTACTS, RESOURCES, AND WEBLINKS

Contacts

Steven Chalk
Acting Program Manager
Solar Energy Technologies Program
DOE Office of Energy Efficiency and Renewable Energy

Craig Cornelius
Solar America Initiative Posture Plan Coordinator
Solar Energy Technologies Program
DOE Office of Energy Efficiency and Renewable Energy

Document Citations

President's Advanced Energy Initiative
www.whitehouse.gov/stateoftheunion/2006/energy

Solar Energy Technologies Multi-Year Program Plan 2007-2011
www.eere.energy

National Energy Policy
May 2002
www.whitehouse.gov/energy

PV Industry Roadmap
www.seia.org

Web Sites of Relevant DOE Offices

Office of Energy Efficiency and Renewable Energy
www.eere.energy

Solar Energy Technologies Program
www.eere.energy

Solar America Initiative
wwwl.eere.energy

Office of Science
Basic Energy Science
www.science.doe.gov/feature/BES.htm

End Notes

[1] 2006 Department of Energy Strategic Plan. United States Department of Energy 2006.

[2] Annual Energy Review 2005. DOE/EIA-0384(2005). Washington, DC: Energy Information Administration.

[3] "U.S. Market Analysis." *PV News* (25:5). Cambridge, MA: Prometheus Institute.

[4]"Solar Energy Technologies Program Multi-Year Program Plan 2007-2011." Office of Energy Efficiency and Renewable Energy. United States Department of Energy. January 2006. pg. 1.

[5] "Forecasts and Analysis- Natural Gas." Energy Information Administration. Retrieved on September 19, 2006 from http://www.eia.doe.gov/oiaf/forecasting. Table 13.

[6] Ibid.

[7] "Natural Gas Summary." Energy Information Administration. Retrieved on September 19, 2006 from http://tonto.eia. doe. govidnavinging_sumisum_dcu_nus_a.htm.

[8] Expected capacity additions for Fossil/Nuclear and Other Renewables for 2005-2030 were obtained from EIA's Annual Energy Outlook (AEO) 2006. To construct the baseline scenario, solar capacity additions for the years 2010, 2015, 2020, and 2025 were taken from data reported under the baseline scenario in the Government Performance and Results Act (GPRA) assessment of EERE's FY07 budget. The baseline solar capacity addition for the year 2030 was taken from AEO's estimate of cumulative solar capacity by 2030. Linear interpolation was used to derive values for interceding years. Solar capacity additions under the low and high scenarios are taken from *SAI Return on Investment* by NREL. It was also assumed that each unit of solar capacity will offset .6 units of other generating capacity, as NREL assumed in the assessment.

[9] Black & Veatch, "Economic Benefits of Concentrating Solar Power in California, Draft Final Report," December, 2005. The quantified potential environmental benefit of solar energy varies depending on the fuel it replaces. The emission reductions were calculated relative to combined-cycle gas plants because a large percentage of the new power plants use natural gas as their energy supply.

[10] "The Solar America Initiative: Energy, Economic, and Environmental Benefits." DRAFT Whitepaper, Roebrt Margolis and the National Renewable Energy Laboratory. September 2006.

[11] "Cumulative installed PV power by application area in IEA reporting countries (MWp)." International Energy Agency. Retrieved on September 29, 2006 from http://www.solarbuzz.com/StatsGrowth.htm

[12] In Spain there are numerous large-scale CSP plants under development and construction. These include the 50MW trough plants Andasol I and Andasol II, and the PS 10, PS20 and Solar Tres power towers. www.mileniosolar.com, www.solucar.es, www.solarpaces.org/News/Projects/Spain.htm. German companies such as FLAGSOL and Millenium AG have played large roles in the design and construction of these plants. See www.mileniosolar.com/, www.flagsol.com. Israeli company Solel has become a major CSP producer/service provider, and is involved with CSP projects in Spain, as well as the SEGS plants in California and a new 64MW plant in Nevada. Solel and the Israeli government are also working together to build a 100-500MW CSP plant in the Negev desert

[13] "Support for PV in Japan and Germany, "Lawrence Berkeley National Laboratory and the Clean Energy Group, 2002.

[14] *U.S. PV Industiy Roadmap Through 2030 and Beyond.* September 2004.

[15] *Solar Annual 2006-The gun has gone off.* Rogol, Michael et al. Solar Verlag GmbH/Photon Consulting. July 2006

[16] *Ibid.*

[17] In Spain there are numerous large-scale CSP plants under development and construction. These include the 50MW trough plants Andasol I and Andasol II, and the PS 10, PS20 and Solar Tres power towers. www.mileniosolar.com, www.solucar.es, www.solarpaces.org/News/Projects/Spain.htm. German companies such as FLAGSOL and Millenium AG have played large roles in the design and construction of these plants. See www.mileniosolar.com/, www.flagsol.com. Israeli company Solel has become a major CSP

producer/service provider, and is involved with CSP projects in Spain, as well as the SEGS plants in California and a new 64MW plant in Nevada. Solel and the Israeli government are also working together to build a 100-500MW CSP plant in the Negev desert

[18] L. Stoddard, J. Abiecunas, and R. O'Connell. "Economic, Energy, and Environmental Benefits of Concentrating Solar Power in California." Black and Veatch. Overland Park, KC. April 2006

[19] Spain introduced a "feed-in-tariff" in September 2002 for CSP-generated electricity and granted a payment of 12 € cents for each kWh output from a CSP plant between 100 kW and 50 MW capacity. In 2004 this was increased under Spanish Royal Decree 436, in which CSP generators receive a tariff of 21 € cents for the first 25 years, and 17 € cents thereafter. Source: IEA Global Renewable Energy Policies and Measures Database: http://wwwlea.org/ textbase/pamsdb/detail.aspx?mode=gr&id
Actual text of Royal Decree (in Spanish): http://noticias.juridicas.com/base_datos/Admin/rd436-2004.html

[20] Stirling Energy Systems Press Release. August 10, 2005. http://www.stirlingenergy.com/ news/SES

[21] San Diego Gas & Electric Press Release, 9/7/2005: http://public Stirling Energy Systems Press Release. September 7, 2005. http://www.stirlingenergy.com/news/SES

[22] "Assessment of Parabolic Trough and Power Tower Solar Technology Cost and Performance Forecasts." Sargent and Lundy Consulting Group LLC. SL-5641. Chicago, II. May 2003.

[23] *Ibid.*

[24] A. Leitner and B. Owens – Platts Research and Consulting/NREL. "Brighter than a Hundred Suns: Solar Power for the Southwest," January 2003. Page 20.

[25] National Renewable Energy Laboratory/Sandia National Laboratories FY2007 AOP. September 25, 2006. Page 28

[26] The stage gate process was originally proposed by Cooper as a model for product development projects and has been extensively modified. See: Cooper, .R.G., Winning at New Products: Accelerating the Process from Idea to Launch" 2nd Edition. 1993, New York: Addison-Wesley Publishing Co.

[27] The U.S. Department of Energy (DOE) administers the Concentrating Solar Power Program through two of its national laboratories — Sandia National Laboratories in Albuquerque, New Mexico, and the National Renewable Energy Laboratory in Golden, Colorado. To increase the administrative efficiency of the program and leverage the respective technical expertise of each of the laboratories, we have combined the concentrating solar power departments of each into a single business unit called SunLab.

[28] An inventory of the Solar Program test and evaluation capabilities, along with references to related codes and standards, can be found at: http://wwwl.eere.energy america/

In: Solar America: How, What and When?
Editor: Nash M. Perales

ISBN: 978-1-60741-333-2

Chapter 3

Solar Energy Grid Integration Systems "SEGIS"

U.S. Department of Energy,
Energy Efficiency and Renewable Energy

1. Executive Summary

The inevitable transformation of the electrical grid to a more distributed generation configuration requires solar system capabilities well beyond simple net-metered, grid-connected approaches. Time-of-use and peak-demand rate structures will require more sophisticated systems designs that integrate energy management and/or energy storage into the system architecture. Controlling power flow into and from the utility grid will be required to ensure grid reliability and power quality. Alternative protection strategies will also be required to accommodate large numbers of distributed energy sources.

This document provides an overview of the R&D needs and describes some pathways to promising solutions. The solutions will, in many cases, require R&D of new components, innovative inverter/controllers, energy management systems, innovative energy storage and a suite of advanced control algorithms, technical methodologies, protocols and the associated communications. It is expected that these solutions will help to push the "advanced integrated system" and "smart grid" evolutionary processes forward in a faster but focused manner.

2. Vision

Solar Energy Grid Integration Systems (SEGIS) concept will be key to achieving high penetration of photovoltaic (PV) systems into the utility grid. Advanced, integrated inverter/controllers will be the enabling technology to maximize the benefits of residential and commercial solar energy systems, both to the systems owners and to the utility distribution network as a whole. The value of the energy provided by these solar systems will increase through advanced communication interfaces and controls, while the reliability of electrical service, both for solar and non-solar customers, will also increase.

3. Program Objective

The objective of this program is to develop the technologies for increasing the penetration of PV into the utility grid while maintaining or improving the power quality and the reliability of the utility grid. Highly integrated, innovative, advanced inverters and associated balance-of-system (BOS) elements for residential and commercial solar energy applications will be the key critical components developed in the effort. Advanced integrated inverters/controllers may incorporate energy management functions and/or may communicate with separate-alone energy management systems as well with utility energy portals, such as smart metering systems. Products will be developed for the utility grid of today, which was designed for one-way power flow, for intermediate grid scenarios, and for the grid of tomorrow, which will seamlessly accommodate two-way power flows as required by wide-scale deployment of solar and other distributed resources.

4. Program Scope

The scope of the SEGIS program includes improving the reliability and increasing the value of PV inverter/controllers while developing interfaces for advanced grid integration. SEGIS products are needed that will increase the value of solar energy systems in today's "one-way" distribution infrastructure and/or will increase the value of systems in tomorrow's two-way" grid or micro-grid. The heart of the SEGIS hardware, the inverter/controller, will manage generation and dispatch of solar energy to maximize value, reliability, and safety. The inverter/controllers will interact with building energy management systems and/or smart loads, with energy storage, and with the electric utility to allow the integration of relatively large amounts of PV energy while maintaining or increasing grid reliability.

Energy management of the future may

- be integrated within inverters or
- be connected via ancillary equipment (portals) that contain the necessary two-way communications to monitor, control and optimize the value of energy produced by PV installations.

Building integration is an important feature of new designs since the complete integration of standardized PV systems with buildings optimizes the building energy balance, improves the economics of the PV system, and provides value added to the consumer and the utility.

The emphasis of the program is on developing inverter/controllers that enable integration of large amounts of PV into the electric utility distribution system.

The scope of the program includes development of inverters/controllers for grid-interactive solar distributed generation systems that either:

- incorporate energy management functions and/or power control and conversion for energy storage, or
- include the ability to interface with energy management and energy storage systems, smart appliances, and utility portals, including adaptation of these systems to communicate with and/or control the inverter/controller.

The following are not within the scope of this program:

- development of photovoltaic modules,
- development of energy storage devices (e.g. batteries),
- non-solar-related development of energy management or energy storage systems, smart appliances, or utility portals.

SEGIS products developed under this program shall be compatible with any of the three primary PV markets segments that are connected to utility distribution systems: residential, small commercial, or commercial. Solar Energy Grid Integration Systems may be configured to address any combination of these market application segments and may be modular in nature. The scale of these markets is described in Table 1.

Table 1. Annlications Scale

Residential	Less than 10-kW, single-phase
Small Commercial	From 10-kW to 50-kW, typically three-phase
Commercial	Greater than 50-kW, three-phase

5. High PV Penetration and the Utility Distribution System

PV systems generate energy with minimal environmental impact. However, a simple PV system without storage provides power only when the sun shines It does not produce power in the evening when loads can be high, and the power output from a PV system can increase or decrease rapidly due to cloud passages. While the markets for grid-connected residential and commercial PV systems are growing rapidly, the total contribution of PV systems to the nation's power supply is small and currently has no significant effect on the operation of the nation's power systems. However, as the quantity of energy generated by solar and other distributed energy systems becomes significant, these systems have the potential to adversely impact utility system operation. To mitigate these impacts, changes are likely to be made to utility/PV system interface requirements and to utility rate structures, which in turn may alter the value of these systems.

a) PV System Characteristics and Impacts

Today's grid-connected residential and commercial systems typically have the following characteristics and associated impacts:

- The PV system and the inverter are connected to the grid in parallel with the load.
  - The load is served whenever the grid is available.
  - Energy produced by the PV system decreases the apparent load. Energy produced in excess of the load flows into the distribution system.
  - The PV system has no storage and cannot serve the load in the absence of the grid.
  - The PV system produces power at unity power factor and utility supplies all Volt Ampere reactive power.

- The inverter meets the requirements of IEEE 1547-2005.
  - There is no direct communication or control between the utility and the inverter.
  - If the inverter senses that utility service has fallen outside set boundaries for voltage and/or frequency or utility service is interrupted, the inverter will disconnect from the utility until normal conditions resume. The load remains connected to the utility.

- For residential and small-commercial systems, the grid interconnection is typically net- metered at a flat rate.
  - The price of energy is constant throughout the day and there is no demand charge.
  - When excess energy is produced, the meter spins backwards.
    - Energy is bought and sold at the same price.
    - Over the course of a month or a year, if energy produced exceeds energy used, the utility will not pay for the excess above the amount used.
  - If the grid is not available, grid-tied PV inverters (without energy storage and load transfer capability) cannot serve the load, even when sunlight is present and the PV modules are able to produce power.

- For large-scale commercial systems, rate structures are more complex.
  - Time-of-use rates often apply, with cost of energy being higher during periods of peak demand.
  - Demand charges may apply with a significant portion of the utility bill derived from the highest power requirement (kW) measured over a 15 to 30 minute interval during the monthly billing period.
  - A charge for VARS (reactive power) may apply.
  - Net metering is less common, and some systems are not permitted to deliver any power back to the utility. In this case, the load must always exceed the energy generated by the solar system.
  - Other systems have dual meters and power is purchased by the utility at a lower rate than the rate charged for power supplied by the utility to the customer.

b) Implications for Utility Operations

There are utility concerns that high penetration of inverter-based solar energy systems along with other distributed generation sources on distribution lines will contribute to instabilities and possibly unsafe operations due to one or several of the following design and operational characteristics:

- Because PV energy production does not always coincide with the times when it is most economical for utilities to use it, it can negatively impact utility operating economics.
  - No power is available in the hours immediately after sunset when demand for power may be high. Thus, the utility must increase peak power generation during these hours.
  - Conversely, utility demand is low in the early morning hours (sunrise to ~ 9 am). Power from solar systems during these hours results in a lower load for the utility, decreasing the need for economical 24-hr base load power, and increasing the need for more expensive intermediate and peaking power during the rest of the day.

- From the utility perspective, net-metered, flat rate customers, especially those whose net demand approaches zero, do not pay a fair share of their costs.
  - If energy generated equals energy used, then energy-related charges (the dominant part of most residential and small commercial bills) will approach zero.
  - Without a demand (kW) charge or significant interconnection charge, customers will pay little for the benefits of being connected to the grid.

- Power production from an individual PV system may increase or decrease rapidly due to cloud passages.
  - In most cases, the rate of change of the collective output from PV systems will be moderated by the geographic dispersal of the systems. However, in a case where the service area is relatively small and rapid weather changes can occur, measurements conducted by Tucson Electric Power show that a rapidly-passing cloud bank can essentially eliminate all solar generation across Tucson in less than 5 minutes.[1]
  - The introduction of significant amounts of rapidly-changing intermittent power in a utility system can affect the controls on and increase the need for spinning reserve.

- If a utility experiences sagging voltage under high demand conditions, IEEE 1547 requires that inverters disconnect. However, since the loads are not automatically disconnected, the utility will see an increase in demand, potentially aggravating the cause of the voltage sag and leading to a blackout (decreased utility system reliability).

- The addition of large numbers of inverters has been shown to increase the probability of islanding, during which inverters continue to supply local loads after a utility fault.[2] Other impacts to the utility include:
    - Inverters are limited in their ability to introduce extremely high levels of short circuit current, but the addition of large systems or many small systems can sum to significant short circuit currents and possibly cause equipment malfunction or damage.
    - Utility protection relays are designed to detect a fault, e.g. an arc to ground created by a tree branch falling across a line. The relays briefly disconnect from the fault to allow the fault to clear, and then reconnect to provide continuing service. If islanding detection fails and inverters remain on-line:
        - The inverters may be damaged by the reconnect.
        - The inverters may continue to supply current which could maintain the fault, causing the utility protection relays to lock open. Utility technicians must then be dispatched to reset the relays, and customers may be left without power for a significant time.
        - When technicians are dispatched, if the inverters are still on-line, the safety of the technicians is threatened because of power being supplied to the load side of disconnects and downed power lines.

- Large power flow into distribution systems that were designed for one-way flow may impact system regulation and protection. Fully understanding the effects of high penetration on distribution system operation and design requires further study, but a number of potential issues have been identified.
    - Reverse power flow can interfere with voltage regulation. Voltage regulators measure current and voltage and will maintain higher voltage at the beginning of a radial line to ensure adequate voltage at the end of the line, in proportion to power flow. Introduction of significant distributed power downstream from the voltage regulation system will make line loading appear to be low.
        - If the power is injected just downstream from voltage regulation, customers at the end of the line will experience low voltage, as shown in Figure 1.
        - If the power is injected near the end of the line, high voltage may occur at that point, as shown in Figure 2.
    - Fuses are designed to protect the current carrying capability of a line, but injection of power downstream from a fuse will not be detected, leading to the potential for overload.
    - Frequency regulation may be impacted when distributed generation from PV systems becomes greater than local conventional generation. The effect on frequency regulation has not yet been observed, but is dependent upon the characteristics of the loads and the load-to-generation ratio.
    - Phase-to-neutral over voltage may develop with load/generation imbalance or with phase to neutral faults. This condition will likely be worsened with high penetrations of PV distributed systems with no dispatchability or interactive controls. Single- phase systems installed disproportionately on a single phase

may cause severely unbalanced networks leading to damage to controls or transformers.

- Inverters using pulse-width-modulation schemes to regulate their outputs typically do not add to lower number harmonics, however the higher frequencies associated with the power electronics do inject higher order harmonics.
  - The levels of higher order harmonics from inverters that would interfere with the distribution equipment have not been fully characterized and need to be studied to determine if there is a possible effect on safety and equipment ratings. However, if required, mitigating the impact by filtering is generally simple, but will impact cost and performance
  - Inverter-generated pulses associated with impedance detection for anti-islanding also accumulate in high penetration scenarios and may cause out-of-spec utility voltage profiles.

Each of these has the potential to decrease system reliability or to increase utility costs in a high-penetration scenario.

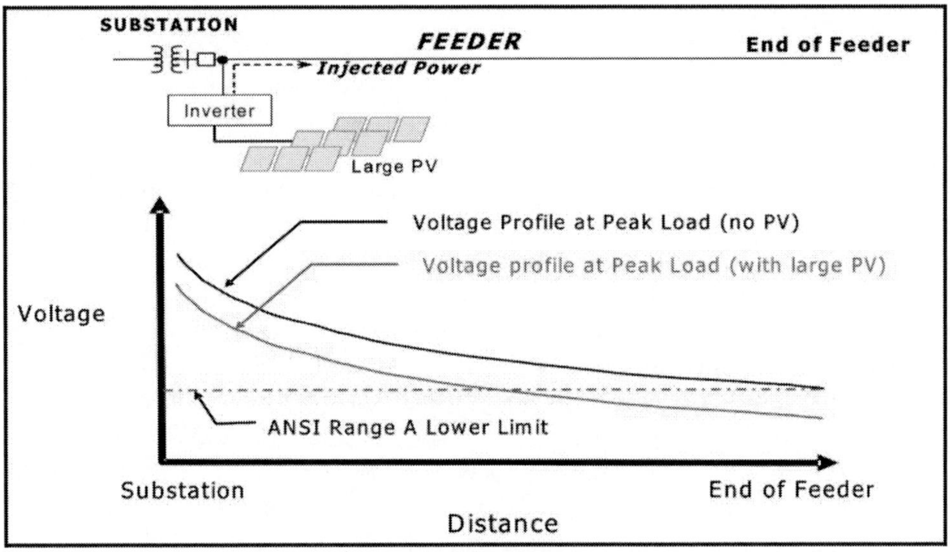

Figure 1. Line Drop Compensation Confused by a PV Generator near Front of Regulation Zone[3].

c) Implications for Solar System Owners

In most cases, authorizing legislation for net metering programs also places caps on enrollment. In the absence of changes to inverter design or distribution system architecture, as the amount of interconnected solar and other distributed generation systems increases, future solar energy customers will likely face alternate metering strategies or limits on supply of power to the electrical grid:

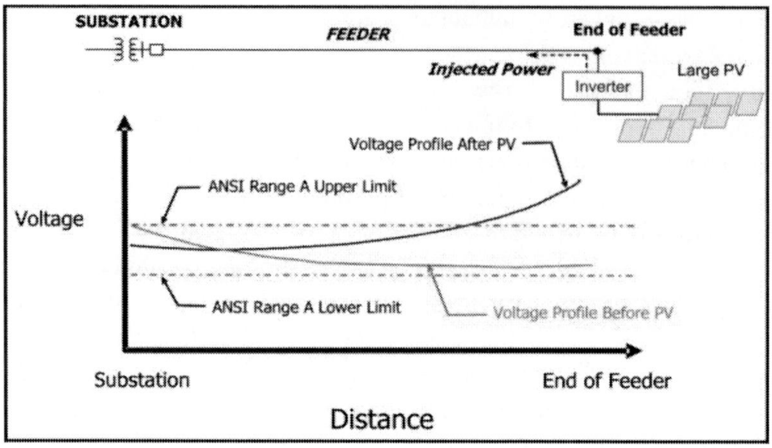

Figure 2. Tail End of Regulation Zone Forced to High Voltage Due to Large Exporting PV System Located Near the End of the Feeder (or Regulation Zone)[3]

- Net meters may be replaced by dual meters, which allow the utility to apply one rate for power supplied to the customer and a lower rate for power provided by the customer. Alternatively, net meters may be replaced by meters that turn only one way, giving no credit for power supplied to the grid. Ultimately, to protect their systems, utilities may need to physically limit power flow into the distribution system.
- Time-of-use rates may be imposed to reflect the true cost of supplying electricity. Time-of-use rates were briefly imposed on new, net-metered, residential solar systems in California. The high rates for power from sunset to 9 pm greatly reduced the value of PV systems, and demand for new systems declined sharply.
- The utility rate structure may be changed by lowering energy charges and imposing infrastructure and demand charges that capture the cost impact of peaking power generation and peak load on distribution systems. In this rate structure, the energy portion of the bill, valued in 0/kWh, decreases and is replaced by a demand charge ($/kW for 15-30 minutes of peak demand) As a result, the value of the energy from a solar system is reduced, and a cloud at the wrong time or demand after sunset can lead to a high demand charges.
- Businesses and industrial sites with large motors are provided Volt-Ampere reactive power (VARs) by the utility. Charges for this capability will likely be imposed on users unless PV or renewable energy generation can supply the VArs at the source.

While decreasing system cost is one approach to providing competitive energy, managing energy flows to improve system value is likely the needed approach for the future grid.

Each of these has the potential to decrease the value to the owner of the energy from the solar energy system. While decreasing system cost is one approach to providing competitive energy, managing energy flows to improve system value is likely the needed approach for the future grid. Designing systems to manage energy flows by incorporating features such as energy management and/or energy storage will allow customers to improve system value by selling energy to the utility when the value is high, buying energy when the cost is low. Such systems also make it possible for the customer to operate independently of the grid, for example during storm-related grid outages. These value-added concepts are discussed further in the following sections.

6. Approaches to Enable High Penetration

a) Today's Distribution System

Utility distribution systems typically are designed for a 30-year life, but in reality they last much longer than that. As a result, in the near term and for most locations, the design of PV systems in a high-penetration scenario must be compatible with reliable operation of a grid infrastructure that will be little changed from current practice. The alternative, saddling individual PV installations with the cost of upgrading the distribution infrastructure, may not be economical. At the same time, it is realistic to expect that most new systems will not be net-metered under flat rate structures, but may be subject to time-of-use, demand, and other charges while potentially receiving less for excess generated energy. In this section, approaches to enabling high penetration of PV into the current distribution infrastructure while maintaining or improving PV system value and utility system reliability are discussed.

i) Mitigating Impact on Current Distribution Infrastructure

As described above, high penetration of PV systems has the potential to interfere with grid reliability and control. Even as utilities contemplate replacement of the existing infrastructure with new hardware and controls specifically designed to accommodate two-way power flows, steps can be taken to mitigate the impact of high penetration of distributed energy sources on the distribution infrastructure. These steps assume that the most expensive parts of the distribution system remain unchanged, but they do require additional communication capabilities.

A key step is to enable the distribution system to interactively control connection of the inverter/controller to the grid. One possible approach is to replace the current anti-islanding approach with interactive dispatch so the utility can command an inverter to ride through voltage sag, rather than having a large number of inverters go off-line while leaving the loads on-line.

Interactive control will also enable the distribution system to direct the inverter to go off-line when there is a fault, rather than relying on multiple inverters, each supplying power to the grid, to independently detect a fault or islanding condition.

Additional steps involve enabling management of power flows to keep high power flows off the distribution system when they could interfere with system regulation and protection, to provide power flows when needed and available, and to mitigate rapid transients. For a centralized power system, the ideal load is predictable and constant, or is large when system demand is low and small when system demand is high. Thus, one approach is to use the SEGIS to manage the power flows by directing loads to operate when solar energy is available, to not operate during peak demand periods when solar energy is not available, and to dampen the impact of transients. The addition of storage would provide more flexible load management. Also, unlike rotating generators with the necessary controls, static power converters (inverters) can be directed to produce reactive power. Inverters can also assist with voltage regulation by supplying VArs and dispatched power.

Alternatively, or additionally, communication capability can be added to allow the distribution system to signal the SEGIS to dispatch power and loads to optimize power flow to the utility. Such communication could be in the form of time-of-use rates and demand charges or could be real-time pricing via smart-metering technology.

ii) Improving Value for the Solar Energy System Customer

As deployment of distributed PV systems increases, many customers are likely to be subject to time-of-use rates and demand charges, and will be paid less for energy delivered to the utility at a particular time of day than they will be charged for energy delivered by the utility at that same time. However, by employing control strategies that optimize value, a SEGIS can mitigate the effects of these rate structures or even exceed the value received by today's net-metered flat-rate customers. The SEGIS will need to manage power flows to and from the utility so that power is purchased from the utility mainly when rates are low, peak demand is minimized, and power is sold to the utility mainly when rates are high.

A first step in optimizing value is to consider the rate structure when establishing the orientation (azimuth and tilt) of the PV modules. At least for regions where utility demand peaks in the summer and time-of-use rates apply, having a portion of the array facing west may provide more value to the customer, even though total energy delivered may be lower.

A second step in optimizing value is to dispatch loads to operate in concert with the availability of solar energy and/or cheap utility power. This could occur by direct communication between the inverter and smart loads via standardized protocols or via inverter communication with an energy management system. Note that some loads may have their own storage and thus are particularly suited to dispatch. For example, tank-type electric water heaters store hot water and can be controlled to heat more water only when either excess solar energy or cheap utility power is available. Even the thermal mass of a building can be used to minimize evening demand, as illustrated by a Florida Solar Energy study of a home with south and west facing PV and a multi- speed air conditioner, shown in Figure 3.[4] This study illustrates the importance of building system design in optimizing solar system value. If PV had been added to the control home, peak daytime demand would have been reduced, but demand after 6 pm would have remained high. On the other hand, if the energy-efficient home had been built without PV, there still would have been significant daytime peak demand (the sum of the red and blue lines). It is the combination of PV and building system design that results in nearly complete elimination of the evening peak. Note that optimizing the performance of a system with more than one orientation for the PV array requires either multiple inverters or inverters with multiple maximum power point tracking inputs.

A third step in optimizing value is to add energy storage so excess solar energy or cheap utility power can be stored for later use when building demands dictate and excess or cheap power is not available. For buildings with demand charges, addition of storage has been shown to add value to the PV system[5], as shown in Figure 4. Storage can have the added benefit of enabling the building to continue to operate critical loads during a utility outage.

To maximize the benefit of the PV system without an undue burden on the building owner/operator may require a control system with adaptive logic. Such a controller would monitor demand, solar energy supply, and utility rates to optimize the flow of energy based on time of day, day of the week, and time of year by controlling dispatchable loads and storage operation. The controller might also communicate with utility smart-metering devices to obtain real-time pricing, and via the interne would monitor weather trends and forecasts to anticipate the availability of the solar resource as well as the real-time price of solar power.

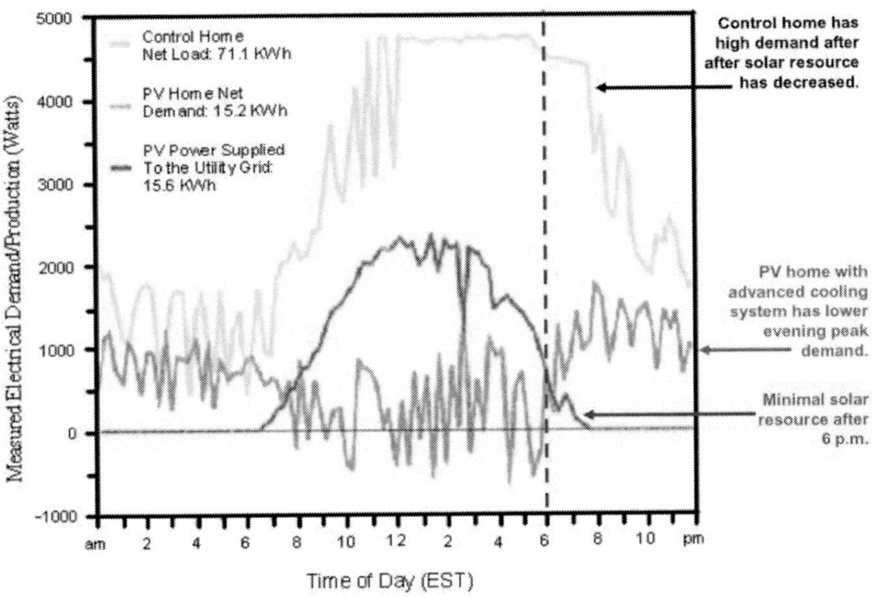

Figure 3. Peak Reduction from Combined Use of Solar Energy and Demand Management in a Residential Application[4]

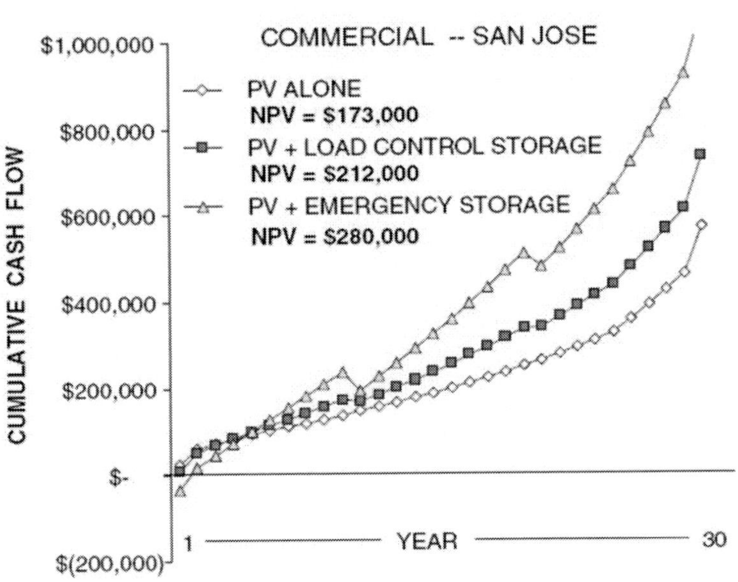

Figure 4. Cumulative Cash Flow for Commercial Customers — Comparing PV-alone, PV + Local Load Management (MBES) Storage and PV + Emergency Storage Options.[5]

b) Advanced Distribution Systems and Micro-Grids

As penetration of distributed energy systems increases and the necessary technology is developed, it is anticipated that the distribution system will evolve to accommodate two-way power flows and will take full advantage of distributed energy resources. This evolution will take place first in new residential, business, and industrial developments, but for much of the grid, may not occur until existing distribution systems must be replaced due to age. The technology to dispatch residential and commercial loads, such as air conditioning and electric water heating, already exists and is being used during periods of peak demand. Customers for which a secure source of electric power is essential use standby generation and/or uninterruptible power supplies to operate independently of the electrical grid during outages, and some use advanced technology to transfer seamlessly from grid power to independent sources. However, with the exception of residential PV systems, flow of power from distributed energy sources into the distribution grid is relatively uncommon.

If the recent growth rate in installation of PV systems continues, penetration levels of 5 to 10% could occur in less than a decade[6], with higher levels occurring locally. Deployment of other distributed resources may also increase. Those resources include, but are not limited to: intermittent resources such as wind and solar; fuel-based energy sources in combined heat and power systems, such as fuel cells, microturbines, and engine-generators; and dispatch of engine-generators that today provide only emergency power in the event of a utility outage. Power flow from these systems may provide a significant portion of the energy used by other customers.

Utilities will use these systems to minimize the cost and environmental impact of power generation, to manage peak demands, to provide ancillary services such as spinning reserve, to improve system reliability, and to improve power quality. At a site level, the ability of power electronics to improve power quality and provide seamless transfer from utility to site power sources has already been demonstrated'. Utilities may place utility-owned distributed generation and/or storage within the distribution system to improve operations and reliability. Micro-grids, placed at critical weak points of the utility grid, can provide a stabilizing role to the grid while ensuring the loads connected to the micro-grids during utility outages continue to function just as if the utility were still there.

To accommodate and optimize management of these resources, it is expected that the advanced distribution system will be a "Smart Grid". One general defmition for the smart grid is:

> A modern distributed and intelligent generation and distribution network using digital technology for controls and advanced electronics for switching and protection.

Features of the smart distribution system may include but are not limited to:

1. Automation for power flow and energy management
2. Management of the interface between the utility, distributed resources, and micro-grids
3. Management of all power flow transitions
4. Real time pricing and analysis for the connected community
5. Management of the intermittency of renewable solar resources

While replacement of the nation's entire distribution system will take many years, advanced distribution system technology and micro-grids will occur as new communities and developments are built and as the technology matures. Micro-grids are especially likely to be deployed where there are critical loads, such as high-technology business parks and/or critical infrastructures such as police and fire departments, hospitals, and water-treatment facilities.

The following are some of the elements that may be found in advanced distribution grids and micro-grids:

- **Electronically Controlled Distribution Systems.** Today's electro-mechanical control systems will be replaced by digital control with new electronic devices providing the controls and likely performing many of the high-speed switching functions.

- **Integrated Electricity and Communication Systems.** The utility distribution system will incorporate communication systems to control distributed generation and storage systems and dispatchable to improve system efficiency and stability while optimizing the value of renewable energy resources such as photovoltaics.

- **Integrated Building Energy Management Systems.** Energy management will be an integrated function when new distributed generation is employed. Building-integrated electrical generation will include energy management systems to optimize the building energy generation value while providing intelligent functions such as load shedding or shifting and energy storage to provide the most value to the consumer and dispatchability for the stability of the utility. Energy management will also likely include intelligent and adaptive logic that manages heating, cooling and lighting needs.

- **Smart, end-use devices.** An integrated system that utilizes communications will enable the electric system to communicate directly with end-use devices, and would automatically optimize system operation.

- **Meters as Two-way Energy Portals.** The meter/service panel for a building will be transformed into an intelligent electronic gateway. Advanced meters/service panels that enable electricity suppliers and customers to communicate in real time and optimize the performance and economics of the system will interact with smart inverters and controllers.

- **Combined Heat and Power:** Distributed generation (DG) systems that enable production of both electricity and processing heat are efficient systems that can augment a central system to improve quality of service.

- **Direct current (DC) in micro-grids.** The concept of power systems that generate and deliver direct current, such as PV/energy storage systems, may be revived to serve DC loads in energy efficient micro-grid infrastructures.

It is evident that communications are a key part of the "smart grid," and standardized communications protocols will be very important to equipment manufacturers of all types, including manufacturers of distribution system hardware and controls, distributed generation equipment, energy management systems, and smart lighting, appliances, and heating and air conditioning systems. Communication needs specific to SEGIS are described in section 7b.

7. Design Concepts for Integrated Inverters, Controllers, BOS and Energy Management

In this section, design concepts are presented for enabling increasing value and reliability for solar energy systems assuming a high level of penetration into the electrical distribution system. A possible scenario for evolution of distributed solar energy systems and the electrical grid is outlined in Figure 5. Currently, PV penetration is small, and most residential customers are net- metered under flat rates. PV inverters detect out-of-spec or loss-of-utility power and automatically disconnect from the grid. Distributed generation systems are locally-controlled and provide combined heat and power, usually with no delivery of excess power to the grid or are used for stand-alone power generation. Some utilities are able to dispatch customer loads, such as air-conditioning and electric hot water heating, during periods of high demand.

PV penetration could grow to as much as 5-10% within the next decade. Initially, accommodating significant penetration of PV and other distributed generation sources will require that most systems be designed within the constraints of the current distribution system. Active utility anti-islanding control of inverters will be required to prevent islanding while permitting ride-through when the utility is stressed. As net-metering caps are met, new PV systems will optimize value to the customer by monitoring rate structures or real-time utility pricing to determine when to operate loads, charge storage, or deliver electricity to the grid. As advanced distribution systems and micro-grids that are able to reliably operate with or without being connected to the utility grid are developed, these systems will be utilized as utility systems are extended to new developments, especially where a secure electricity supply is essential. The availability of plug-in hybrid vehicles may significantly change the demand profile of residential and small commercial customers.

As deployment of distributed generation increases and existing distribution infrastructures must be replaced due to age and obsolescence, additional features will be added, including management of substantial distributed energy flows into the utility grid via utility dispatch of distributed generation and storage resources, digital electronic control of the distribution system, and increased use of utility-interactive building energy management systems and smart loads, and micro-grids. As distributed generation technologies mature, DG will likely provide much of the capacity additions to the nation's electrical system. As fossil fuel prices increase, use of electricity for transportation may also increase both directly and via use to produce alternative fuels.

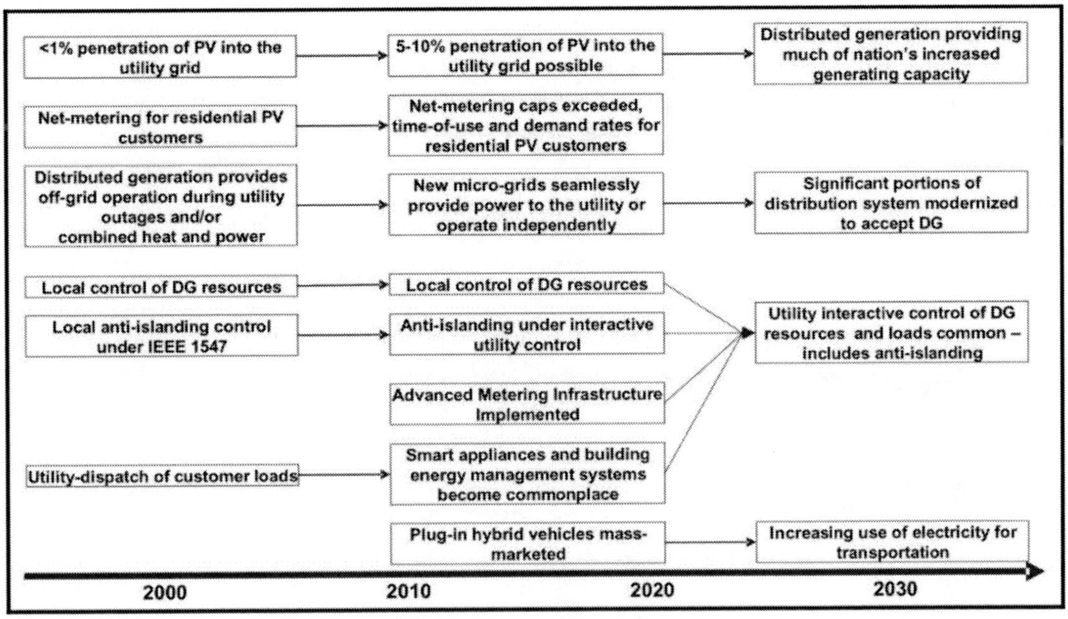

Figure 5. Possible Time Tine for Introduction of Distributed Energy Resources and Advanced Infrastructure

a) System Architectures

Integrated power conversion topologies for distributed generators are to be developed through the SEGIS Program with new emphasis on energy management and grid integration. The advanced inverter/controller topologies may include an energy storage element or load control when intended for stability of the energy supply. The stored energy component should be remotely dispatchable or contain built-in intelligence. Intelligence will enable the capability to characterize the integration interface and to adapt to anticipated and real changes related to the state of the utility grid (i.e. does the grid require assistance from the renewable energy source or should local loads be manipulated to maximize the benefits).

Figure 6 shows the topology of a typical net-metered PV system, in which power is supplied to the grid, when available, and the inverter monitors the grid as required by the IEEE 1547 standard. No communication occurs between the system and the grid.

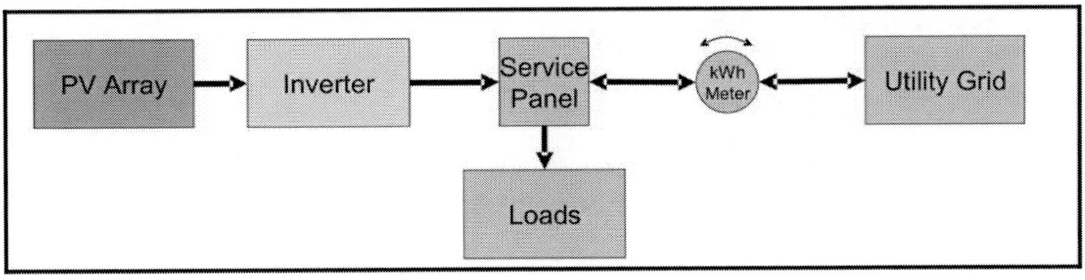

Figure 6. Today's Net-Metered System

Figure 7 shows a system with storage. The primary purpose of storage in this system configuration is to serve critical loads during a utility outage. During an outage, the inverter disconnects from the main service panel and the utility and serves critical loads via a sub-panel. If the system is operated under a time-of-use rate structure, it is also possible to operate the system to increase value by charging storage with utility power when utility rates are low and discharging storage to the utility or to the loads when utility rates are high[13].

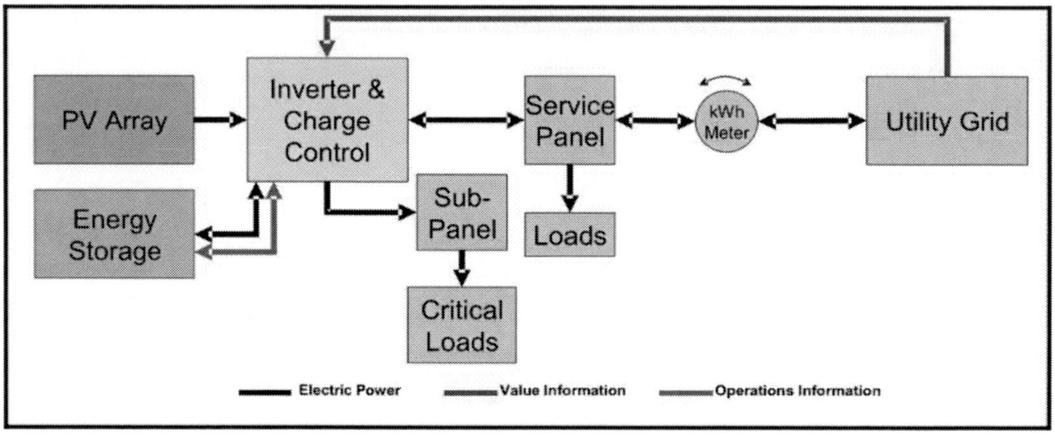

Figure 7. Today's System with Storage

The next two configurations, shown in Figures 8 and 9, are designed for integration with today's distribution system. However, they include enhanced communication to enable utility interactive anti-islanding control and to provide value information, such as real-time pricing. No other provisions are made for direct control of power flow by the utility. The system control function shown in the figures represents the heart of the SEGIS. The functions shown: inverter, energy management, adaptive logic, and, in Figure 9, charge control, could all be incorporated into a single device, or could be separate devices designed to communicate with each other.

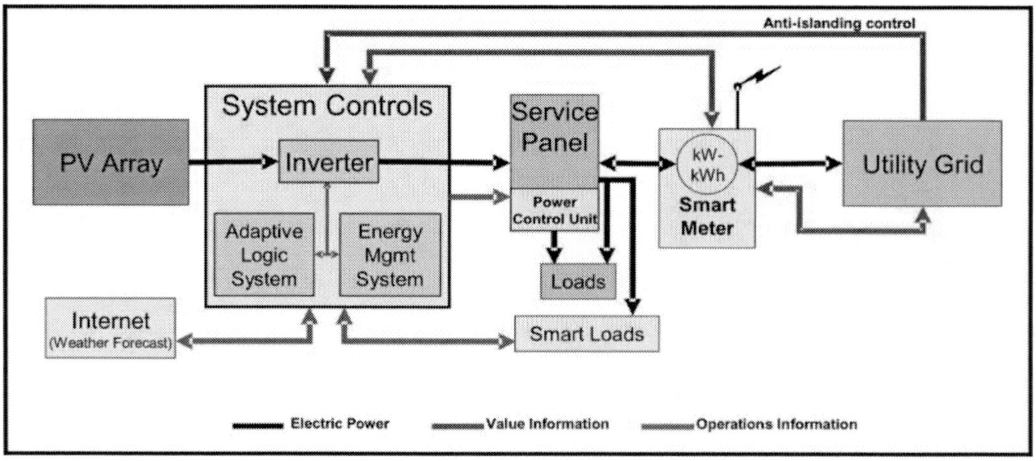

Figure 8. Advanced System without Storage

The system in Figure 8 optimizes value and manages power flow to and from the utility by controlling loads. The smart metering system may provide real-time pricing or may simply represent a set rate structure (time-of-use and demand charges) which would be programmed into the Solar Energy Grid Integration System. Knowledge of utility rates enables the SEGIS to control the loads and associated energy flow to and from the grid to provide increased value to the customer. Smart loads are controlled by an information signal from the Energy Management System. A power control unit is used to remove power from loads which are desirable to control, but do not incorporate information links. An optional adaptive logic controller enables the system to monitor loads, utility rates, and, via the internet, weather conditions and forecasts to provide optimum control of energy flows to maximize value to the customer. The adaptive logic controller could also monitor expected system output as a function of solar input and would signal the need for maintenance to address out-of-nominal operation.

Adding storage to the system, as shown in Figure 9, has the potential to further increase the value of the system. In the figure, the charge controller that supplies energy to the energy storage unit and the inverter that converts the output of energy storage to AC power are shown to be part of the PV inverter/controller. However, this is just one of many possible configurations. DC chargers, AC chargers, and dedicated inverters for storage are all available and could be operated in coordination with the PV inverter. While PV can charge storage via a DC charger, an AC charger may be preferred since at least some of the charging will be done when utility-supplied electricity is lower cost (overnight for a summer peaking utility).

A desirable feature for a system with storage is to enable the system to operate independently during a grid outage. Usually, to maximize the period of independent operation, essential and non-essential loads are separated at the load center and/or by the EMS, and only essential loads are supported with energy from storage.

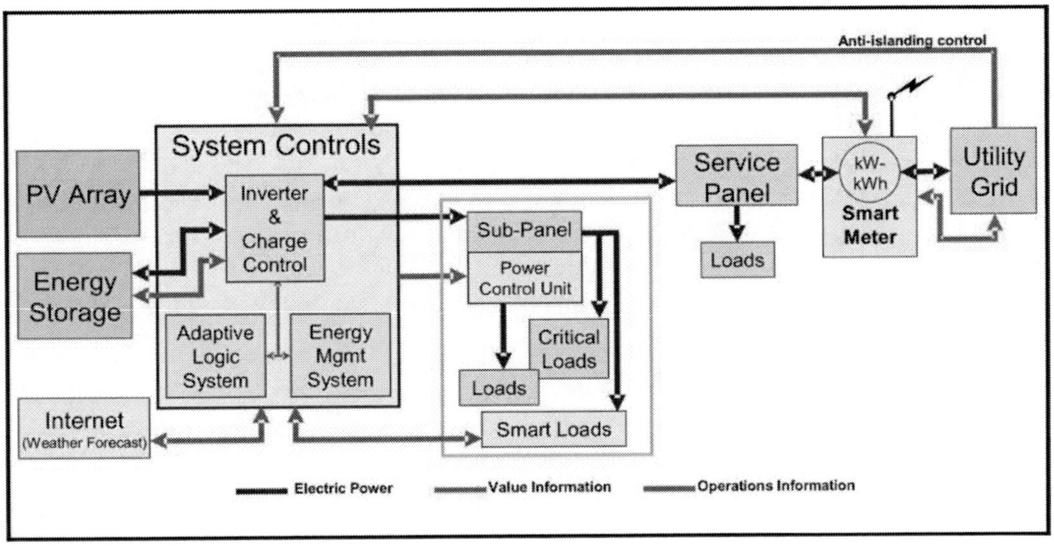

Figure 9. Advanced System with Storage

In Figure 10, provision is made for control of the distributed energy system by the advanced distribution infrastructure via a portal associated with the smart metering system. All communication flows through this portal, including the ability to intentionally dispatch energy from the system or to request the system to operate independently of the grid. Two-way communication is also shown so that the SEGIS systems are able to report their status, including the availability of solar or stored energy, to the utility.

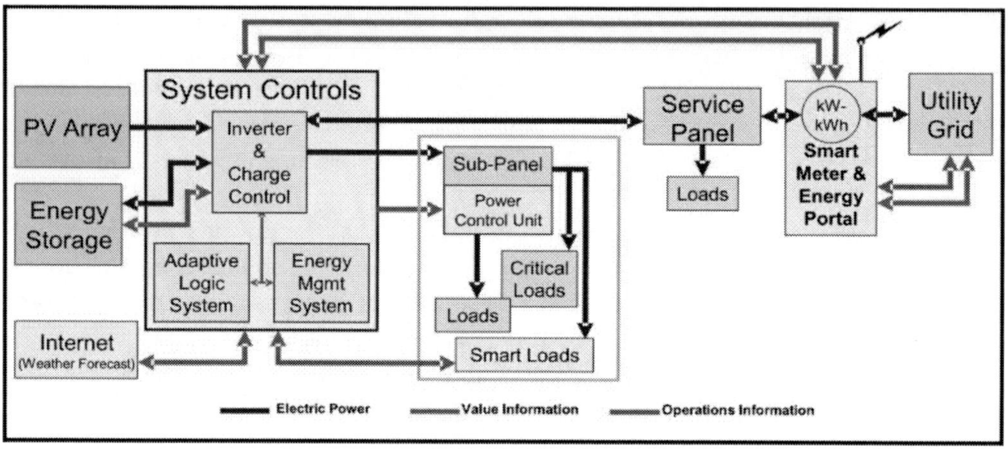

Figure 10. System for Advanced Distribution Infrastructure

In a variation of this approach, a micro-grid might have multiple energy sources, storage systems, and building load centers tied to a single controller that will provide supervisory control of the entire micro-grid system, especially during operation independent of the utility.

The above figures illustrate a few potential system configurations. Many others are possible. Among the many factors to be considered in optimizing system architecture are the number of energy conversion steps in converting DC and AC power from the PV array, round-trip efficiency of the energy storage, the optimum voltages for the PV array and the battery bank, maximizing system value during grid-tied operation vs. maximizing the capability for grid-independent operation, storage capacity to shift loads relative to time-of-use rates or to avoid demand charges, storage capacity vs. depth of discharge to optimize system life and charging efficiency, levels of redundancy vs. common points-of-failure that may undermine reliability, impacts of PV intermittency, etc. Key goals of the SEGIS program are to encourage innovation in system-integrated inverter/controller products, total system reliability/architecture and system integration.

b) Communications

Communication is a critical function for the Solar Energy Grid Integration System. As PV systems increase in number and penetration, communication with the distribution system operator will be essential to ensuring safe, reliable operation. Other communication functions will be critical to optimizing system value. To enable widespread use both by the ~3,000 utilities in the U.S. as well as by the building industry, SEGIS should use protocols based on open system standards in order to offer the broadest applicability. Additional standards development may be necessary, for example, to support anti-islanding control. Developing communications capabilities that are compatible with other national standards, such as in Europe, will increase the marketability of SEGIS. Providing secure communications between system monitor and control functions spread out over long distances back to a centrally located SCADA (Supervisory Control And Data Acquisition) type administrative control site will increase SEGIS acceptability. Dedicated Virtual Private Networks (VPN) can be implemented to specify security features and separate different user traffic over a SCADA network

i) Anti-islanding Control

The inverters in today's PV systems monitor grid parameters and disconnect when those parameters fall outside the ranges established in IEEE 1547. However, as noted above, as the number of interconnected inverters increases, a more interactive method is needed to ensure that inverters will disconnect when required, but will be able to ride-through variation in utility operating parameters, for example, caused by high demand. Communication of the need to disconnect must be fast (less than one second) and certain (always occurs when connection to the utility is lost). The distance over which this information must travel is potentially long, but the amount of information required is small.

ii) External Communication

To optimize system value, the system must respond to external data, such as utility rate structures, real-time pricing, secure dispatch signals, and weather forecasts. Security is important for both dispatch and anti-islanding signals, because a cyber attack resulting in simultaneous disconnect of a significant number of systems representing significant penetration of a portion of the grid could overwhelm the ability of the utility reserve capacity, causing an outage[8]. Relative to islanding control, the speed with which data must be transferred is lower, depending on the need, as follows[9]:

- **Spinning reserve** (replacement of lost capacity), signal sent ~1/second. Thermal generation units providing spinning reserve achieve full output in a few minutes. Hydroelectric units respond faster.

- **Frequency and Area Regulation** (to maintain frequency control), signals sent very few seconds. Response time is over several minutes.

- **Voltage regulation.** The line voltage provides the signal. Very fast response (sub-cycle), e.g. using droop algorithms is possible. Central dispatch is slower (few cycles).

- **Peak Shaving (Demand Response).** The signal can be built into peak demand rate structures or real-time pricing, which may be updated at intervals of 1 to 60 minutes, depending on the utility. Some utilities also dispatch loads to shave peak demand.

- **Backfeed Control.** Some systems, especially large commercial systems, may be restricted from exporting power to the utility. Often, these systems are sized so that the system output never exceeds load, but in the event it does or if the system is larger and uses storage to absorb excess power, then the response must be faster than the reverse- power relay, which is typically sub-cycle.

iii) Internal Communication

Communication within the system is critical to controlling loads and storage to optimize system value while maintaining system safety. Distance over which data must travel will generally be modest, unless the site is a large commercial system or micro-grid. In addition to ensuring safe, efficient system operation, a system that is connected under time-of-day and/or demand rate structures will be trying to balance the available energy from the solar system, other connected distributed generation sources, and storage with variations in loads and utility pricing. While solar output can change rapidly with cloud passages and loads can change with the flick of a switch, utility demand charges are usually assessed over a period of 15-30 minutes, so the system has time to respond. However, in the event of off-grid operation, the system must act quickly to prevent load from exceeding maximum system capability (maximum inverter rating and/or available solar plus storage system discharge capability) at any given time. If the system is not permitted to export power, then loads, storage, or the inverter output must be controlled to avoid backfeed (less than one cycle).

iv) Communication Methods and Protocols

Newmiller et al. (2007) have provided a summary of the characteristics of a variety of communications methods and protocol. Those and others are summarized here:

- **Dedicated copper wiring.** Large transmission-connected generators use dedicated copper wiring for control, but the cost to connect with large numbers of individual distributed generation systems would be very high.
- Ad-hoc mesh networks. Smart, granular network topologies that utilize ad-hoc connection methods, whereby individual devices discover others within range to form a cooperative mesh communication network capable of establishing a massive infrastructure with end-to-end routing links
- **Continuous-carrier power line communications carriers (PLCC).** Used by some utilities for automated meter reading systems. The PLCC signal is lost if the connection to the utility is lost, so this signal could be effective for inverter anti-islanding control. Drawbacks are high cost, low bandwidth, and high power demand.
- **Broadband-over-Power-Line.** This approach takes advantage of existing power line infrastructure to communicate data, but this technology has faced technical issues, including interference with other radio spectrum users and interference from loads. Some systems have gone to digital multi-carrier modulation scheme to mitigate radio interference issues.
- **Ethernet.** Provides communications within buildings, but must be connected to a widearea-network technology, such as cable television. Wide-area networks do not have sufficient reliability to support protection functions.
- **Wireless Local Area Network (IEEE 802.11).** Used inside and outside of buildings to provide short distance wireless data.
- Wireless Interoperability for Microwave Access (WiMAX) (IEEE 802.16) Provides longer range wireless access.

- **Wireless Metropolitan Area Networks (WirelessMAN or WiMAX, IEEE 802.16d).** A standards-based technology enabling the delivery of *last mile* wireless broadband access as an alternative to cable and DSL. Used in urban environments to transmit 2km without line-of-sight antenna configurations and up to 10km with unobstructed path.
- **Personal Area Networks,** e.g. IEEE 802.15.4, provide short distance (few meters) wireless communications.
- **Energy Management Systems,** BACnet is a standard in America, Europe, and over 30 countries developed by the American Society of Heating, Refrigerating and Air-Conditioning Engineers (ASHRAE-135) for building automation and control networks. ANSI/CEA-709.1-B establishes another standard for control networking. It provides one of the data link/physical layers of BACnet.
- **Other.** A variety of other technologies are available, including conventional telephone land-lines, cellular telephone, spread-spectrum wireless radios and pagers. Some utilities have used wireless technology to dispatch customer loads, e.g. Nevada Power's Cool Share program.

c) Inverter/Controller

Development of the inverter/controller is a central element of the SEGIS program. The inverter transforms dc power from the PV array to grid-quality ac power. Depending upon the system architecture, the inverter may also charge and discharge energy storage, and may control smart loads, e.g. smart appliances, especially in residential systems. The inverter/controller, if it contains adaptive logic, may also determine when excess energy is dispatched to the grid or stored. Commercial systems are more likely to use a separate energy management system with which the inverter must communicate.

A number of elements have been identified for research and development and possible inclusion in the inverter and other elements of the Solar Energy Grid Integration System:

- Integrated circuitries (such as monolithic electronic modules), layouts and packages to improve applications flexibility, facilitate thermal management, survive high-temperature operation, and to ease building integrations.
- Thermal management through innovative technology (e.g. MicroElectroMechanical Systems (MEMS) or innovative components and engineering.
- Engineered transient over-voltage surge protection on both the AC and DC sides utilizing new devices and complete system designs that minimize inductive coupling and induced surges. DC-side over-voltage suppression devices must be capable of clamping transients of up to 300 V/module and 100 V/sec.
- Open communication protocols and determination of the optimal inverter/controller "intelligence" for PV applications.
- Elimination of least reliable components (currently electrolytic capacitors) or the use of selective redundancy that results in low-cost reliability improvements.

- Utilization of new but proven state-of-the-art devices (such as wide-band-gap devices[10], new long-lived capacitors, advanced low-loss magnetics, and innovative packaging and circuitry layouts).
- Reduced cost and complexity of installation through innovative installation methods, reduction in DC-side component count, and reductions in on-site programming and engineering requirements (Plug-n-Play goals).
- Innovative integration of electronics/control and communications with PV modules or within listed PV module packages.
- Self diagnostics for the inverter/controller/energy management and the system with reporting capabilities for remaining lifetime predictions.
- Micro-grid-ready controls that can enable intelligent electrical grid support such as dispatchable or intentional islanding.
- Integrated energy storage control and optimization to maximize the benefits of the renewable energy resource.
- Integration with residential and commercial building energy systems for the retrofit and the new buildings market (e.g. Zero Energy Homes)
- Smart, integrated system controls with algorithms and secure communication capabilities to optimize system value and energy efficiency.
- Compliance with "National Electrical" and "Building" codes and domestic and international standards and certifications.

This list is neither exhaustive nor are all of these elements required. Rather, it is intended to encourage innovation in developing advanced inverters/controllers that will be highly reliable and will add value to the system, beyond just converting dc power to ac power.

d) Energy Management Systems (EMS)

As shown in Figures 8-10, the energy management function is a key SEGIS element. The traditional role is to optimize building energy by controlling energy-using equipment (lighting, air conditioning, etc.) relative to building use (time-of-day/week), weather conditions (available daylight...) and utility rates (time-of-use and demand charges). The addition of an intermittent solar energy resource adds a new and complex variable into the energy management equation.

Energy Management Systems (EMS) are common in large commercial buildings, but typically have not been partnered with distributed generation. For residential and small commercial systems, it is possible that energy management functions may be incorporated into the inverter/controller and/or some of the energy management elements may be incorporated in smart appliances. For larger systems, it is more likely that energy management and the inverter/controller will be separate SEGIS sub-systems. To permit the SEGIS to optimize system value will require development of control and secure communication interfaces between the various sub-systems and may require development of additional capability within the EMS.

e) Adaptive Logic Controller

Controlling loads and managing the flow of available energy from the solar system, the flow of energy to and from the grid, to and from storage, and to loads for the purpose of maximizing system value is not a trivial task. Both the solar resource and utility demand, including real-time pricing, is a function of weather as well as the time of day, day of the week, and day of the year. To avoid the need for power during high time-of-use rates or having high demand for just 30 minutes a month that can significantly increase the utility bill requires anticipation of the solar resource relative to the level of energy stored and likely future energy needs. An adaptive logic controller that also monitors weather forecasts could increase system value by managing these power flows. An adaptive logic controller has the ability to learn. Thus, instead of relying only on pre-programmed energy management strategies, it develops its own optimization algorithms from past energy demand, utility pricing signals, and solar resource availability relative to weather patterns, time of day, etc. Such a controller could be part of the inverter or might be a separate unit that can communicate with the inverter, the smart meter/energy portal, and the internet.

f) Related Systems

There are a number of related systems that must be considered in the design of a SEGIS. The following are among these related systems:

i) **Energy Storage.** As indicated above, energy storage is a key element of advanced power flow management. A relatively small amount of energy storage can be used to minimize the rate of change of system output to the grid and avoid demand charges. Much more storage is required to provide significant capability for off-grid operation, unless other generation sources, such as an engine-generator, are available. Control functions and charging/discharging power electronics may be included in the inverter/controller, as shown in Figures 9 and 10, or they may be independent SEGIS subsystems that can communications with the other SEGIS sub-systems. Understanding the charge/discharge efficiency, maintenance requirements, lifetime, and capital cost will be essential to design of a cost-effective distributed energy system that includes storage.

ii) **Advanced Distribution Systems/Micro-grids.** The architecture of advanced utility distribution systems will influence the SEGIS design, but distribution system development is beyond the scope of the current program. However, SEGIS developers are expected to consider the likely evolution of grid systems in their designs, and, within the scope of this program, may design SEGIS systems with the ability to function as a micro-grid.

iii) **Hybrid Vehicle Systems.** The development and introduction of plug-in hybrid electric vehicle (PHEV) systems raises a number of interesting possibilities in SEGIS design. At a minimum, the PHEV represents a dispatchable load that can be charged when excess or cheap energy is available. However, unlike other loads, the PHEV presents the alternative of self-power which may, at times, be more economical than charging the PHEV. Potentially, the PHEV is also a power source or source of stored energy, but one which may or may not be on-site at any given time. In any case, the PHEV presents a new alternative in power management.

8. BENEFIT/VALUE ANALYSIS

Penetration of PV into the distribution system will accelerate as the cost of energy from PV systems declines relative to conventional sources and the value of that energy improves. While cost of energy is a function primarily of system performance and life-cycle cost, the value of the energy depends on many factors, such as when it is available and the reliability of the energy. Some of the elements that factor into the value of a Solar Energy Grid Integration System are described in this section.

a) Reducing Inverter Cost

The cost of inverters for simple net-metered systems is declining as production volumes increase. However, as penetration of PV into the distribution systems increases and rate structures change, the inverter and related-subsystems must become more complex in order to increase the value of energy relative to a simple, net-metered system. Thus, within the SEGIS program, there will be a need to use some of the advanced approaches outlined at the end of Section 6 to hold costs down even as the systems become more complex.

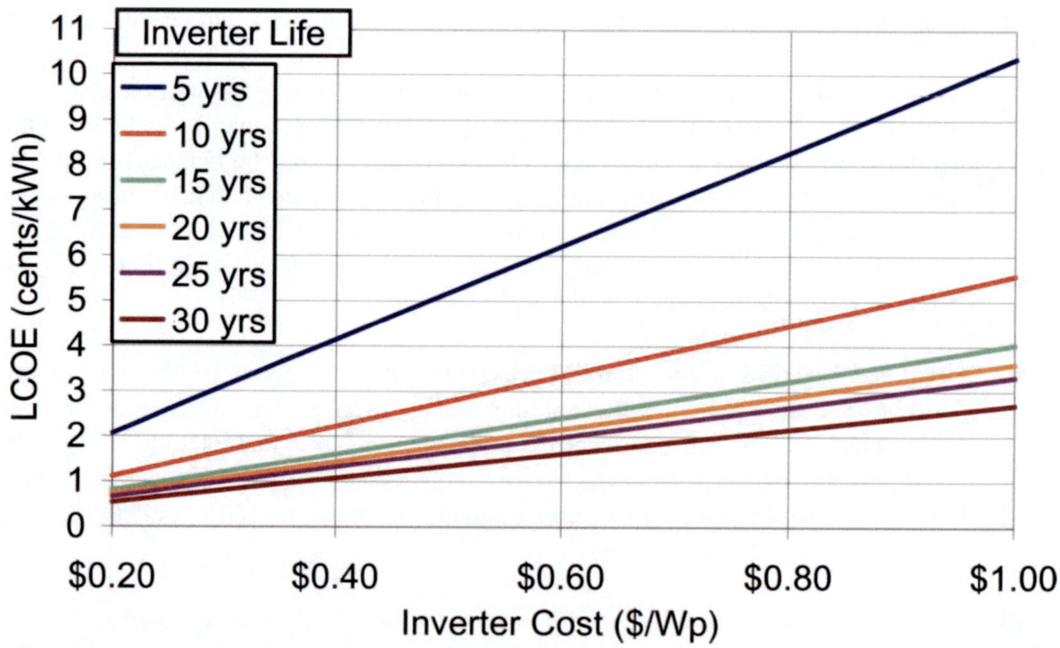

Figure 11. Contribution of Inverter Cost and Replacement to Energy Cost

System reliability is also important to cost reduction, especially for residential systems. For example, a 2-kW residential system in a sunny location will produce about 4000 kWh/yr. If that energy has a value of 10¢/kWh, then the value of the energy from the system is $400/yr. It is easy to see that a $100 annual service call would significantly reduce the value of the system. If an inverter cannot be repaired, but must be replaced, then the impact on system value is even greater. As shown in Figure 10, there is a tradeoff between system cost and system lifetime. Using standard economic assumptions from the DOE solar Multi-Year Program Plan[11] for interest, inflation and tax rates, an inverter on a residential system in Phoenix with a 30 year lifetime and a cost of $1.00/Wp contributes 2.7¢/kWh to the cost of energy. In contrast, an inverter which lasted only ten years and thus had to be replaced twice over the life of a system will also contribute 2.7¢/kWh to the cost of energy if the inverter cost is only $0.50/Wp.[12] If an inverter is to contribute no more than 1¢/kWh to the cost of energy in a sunny location like Phoenix, then the cost of the inverter must be less than $0.40/Wp if it lasts the life of the system (thirty years). If the inverter has a shorter lifetime, then the cost must be even less. As SEGIS inverter/controllers become more capable and complex to improve system value, reducing first cost and improving reliability will both be critical to achieving higher value.

b) Improving System Value

As noted above, a shift away from net-metering to time-of-use rate structures with demand charges has the potential to reduce the value of the energy from PV systems. In a simple analysis of rate structures in Phoenix, the value of energy from a south-facing net-metered PV system is 7.3¢/kWh on the hottest day of the year. In contrast, under a time-of-use rate structure that has a lower cost-of-energy charge (¢/kWh) but adds a peak demand charge ($/kW), the value of the solar energy is reduced to 4.8¢/kWh[12]. Because demand peaks late in the day as the output of the solar system is declining, the solar system contributes little to peak demand, and the demand charge is not reduced. However, as shown in Figure 3 and discussed in section 6, an innovative design strategy can substantially reduce peak demand, even without storage. This is just one example of the many building, energy management, and system design strategies that can be used to improve system value.

c) Value of Storage

Energy storage can be used to minimize peak demand and to shift delivery and consumption of power to minimize energy costs. In the simple single-day analysis cited above for Phoenix, the addition of storage to eliminate peak demand and shift demand to periods when utility rates are lowest has the potential of increasing the value of the energy from the PV system to 9.90/kWh. Of course, there are additional costs as well as increased losses associated with charging and discharging energy storage. Other papers have performed more comprehensive analyses. Stevens et al.[13] show the benefit of using low-cost overnight power to charge energy storage. Hoff et al5 showed that energy storage can be used to provide local load control and to improve value relative to implementation of either PV or storage individually. Further, the impact of rapid changes in sunlight on utility operations in some scenarios can be substantially mitigated with just fifteen minutes of storage capacity.[1] Note that storage options are not limited to electric storage. Among load-side storage options are ice-based air conditioning systems and electric hot water heating. In any case, the capital cost, efficiency, lifetime, and maintenance costs of energy storage must be considered in the analysis and system design.

d) Models for Characterization of Performance and Cost of Energy

Fully evaluating the benefits of a SEGIS system requires modeling of solar energy production, building loads, and energy storage relative to capital cost, maintenance, and the real-time cost of alternate energy sources (utility power). At this time, no single model will perform this analysis, but there are several that can be helpful. The Solar Advisor Model[14] is used to conduct parametric analysis of system designs and to calculate performance and levelized cost of energy. Planned improvements to the Solar Advisor Model will support analysis of the value of systems with storage in a time-of-use rate structure. HOMER[15] can be used to analyze solar energy production and energy storage. PVDesignPro [16] is a commonly-used commercial model for analysis of solar system designs. Building energy analysis models, such as BEOpt[17] and Energy-10[18] may be helpful in understanding building energy demand and use of energy management systems.

9. Conclusions

The SEGIS program is intended to provide the impetus for improving the methodologies and hardware for increasing the penetration of PV systems into the utility grid. The development of advanced, integrated inverter/controllers and associated energy management functions is a critical part of the SEGIS program. The SEGIS program is an aggressive effort to enable substantial penetration of PV into today's grid, into intermediate grid scenarios, and into the smart grid of the future, which will be characterized by a significantly larger amount of distributed generation, much of it from intermittent sources. To achieve optimum value and to enhance the reliability of power for solar systems owners and the grid as a whole, these systems will require advanced controls that can integrate energy management and energy storage.

Changes to the protection methods for the integrated grid will be required to accommodate the possible back-feeding of energy from these sources, and distributed generation systems will be required to demonstrate value in a market dominated by time-of-use tariffs.

Advanced inverter/controllers and energy management systems will need to include sophisticated interfaces and controls to be able to integrate with emerging "Smart Grid" technology, and as such, must be compatible with communication protocols utilized by established and emerging energy management and utility distribution level communication systems. Finally, these systems must meet the performance and reliability targets consistent with achieving levelized cost of energy that will be competitive in future energy markets.

BIBLIOGRAPHY

1. *2008 National Electrical Code*, ANSI/NFPA70, Published by the National Fire Protection Association, Quincy, MA, 2007.
2. King et al, "Performance Model for Grid-connected Photovoltaic Inverters," Sandia National Laboratories Report, SAND2007-5036, September 2007.
3. DOE Advanced Integrated Inverter and Energy Management – Technical Workshop Presentations, April 19, and May 10-11, 2007, Denver, CO and Santa Ana Pueblo, NM http://www2.eere.energy.gov/solar/solar_america/technology_pathway partnershipswks hp.html
4. Summary Report on the DOE High-tech Inverter Workshop, Jan, 2005, http://www1.eere.energy.gov/solar/pdfs/inverter_II_workshop.pdf
5. Galvin Electricity Initiative, "The Path to Perfect Power – The Perfect Power System," 2007, http://www.galvinpower.org/galvinpower.pdf
6. *2005 NEC Handbook*, NFPA70, Published by the National Fire Protection Association, Quincy, MA, 2005.
7. Ton, D., Bulawka, A. & Bower, W. (2003). *Summary Report on the DOE Workshop On a Systems-driven Approach To Inverter Research and Development, U.S. Department of Energy, EERE Solar Technology Program*, http://www1.eere. energy.gov/ solar/ pdfs/sda_inverter.pdf Sep.
8. Bower, W., Ropp, M. (2002). "Evaluation of Islanding Detection Methods for Photovoltaic Utility-interactive Power Systems", *International Energy Agency Photovoltaic Power Systems Implementing Agreement, Task V; Grid Interconnection of Building-integrated and Other Dispersed Photovoltaic Power Systems, Report IEA PVPS T5-09*: 2002, Feb.
9. Lee, F. C., Van Wyk, J. D., Boroyevich, D., Jahns, T., Lorenz, R. D., Chow, T. P., Gutmann, R. J. & Barbosa, P. (2002). *An Integrated Approach to Power Electronics Systems*, Report for National Science Foundation.
10. *UL Standard for Safety for Static Converters and Charge Controllers for Use in Photovoltaic Power Systems, UL1741,* Underwriters Laboratories, Second Edition, Jan 2001.
11. IEC 61683. (1999). *Photovoltaic systems - Power conditioners - Procedure for measuring efficiency*, International Electrotechnical Commission, Pub. Nov.
12. *Beyond the "Balance of System - Growing Value of Solar Installations*", Presented at Solar Power 2006, San Jose, CA, Sep 2006.
13. ISET – Advancing Energy Systems, 2006 Annual Report, Institute for Solare Energieversorgungstechnik, Kassel, Germany

REFERENCES

[1] Henry, B. (2007). *"Utility Needs for Advanced Intelligent Inverters,"* Proceedings of the DOE Advanced Integrated Inverter and Energy Management – Technical Workshop, Santa Ana Pueblo, NM, May 10-11, http://www2.eere.energy america/pdfs/8_bill_henry_presentation_051007.pdf .

[2] Horgan, S. (2004). *"Distributed Utility Integration Test (DUIT): Facility and Unintentional Islanding Results,"* Presented at the US DOE High-tech Inverter Workshop, April.

[3] Barker et al. (2007). *"Renewable Systems Integration: Report 1, Advanced Grid Planning and Operation,"* Electric Power Research Institute, Draft October.

[4] USDOE, U.S. Department of Energy (1999). *Making the Most of Residential Photovoltaic Systems*, DOE/GO 10099-918, September.

[5] Hoff et al. (2005). "Maximizing the Value of Customer-sited PV Systems Using Storage and Controls," *Proceedings of the 2005 Solar World Congress*, Orlando, FL, Aug.

[6] Newmiller, et.al. (2007). "Distributed PV Systems Design and Technology Requirements," RSI Study Report, BEW Engineering, October

[7] S&C Electric Company, PureWave UPS System, http://www.sandc.com/products/ purewave/default.asp

[8] McIntyre, A. (2007). *"Renewable Systems Interconnection Security Analysis,"* Sandia National Laboratories, draft, September.

[9] Newmiller, et al. (2007). "*Distributed PV Systems Design and Technology Requirements,"* BEW Engineering, Draft, October.

[10]Burger et al. (2007). "*Photovoltaic Inverters with SiC MOSFETs,"* Fraunhofer Institute for Solare Energy Systems (ISE), Freiburg, Germany, presented at 2 nd ECPE SiC User Forum, Copenhagen, September 6-7,.

[11] *"Solar Energy Technologies Program Multi-Year Technical Plan,* 2007-2011," U.S. Department of Energy, http://www1.eere.energy.gov/solar/pdfs/set_myp_2007-2011_proof_1.pdf

[12]Cameron, C. P. (2007*). "Impacts of Advanced Inverter Lifetimes and Reliability on Levelized Cost of Energy,"* Proceedings of the DOE Advanced Integrated Inverter and Energy Management – Technical Workshop, Santa Ana Pueblo, NM, May 10-11, http://www2.eere.energy.gov/solar/solar_america/technology_pathway_partner ships_wkshp.html

[13]Stevens, J. W. et al, "Performance of a grid connected residential photovoltaic system with energy storage", SAND97-2742C, http://infoserve.sandia.gov/sand_doc/ 1997/972742c.pdf

[14]Solar Advisor Model (SAM), NREL, http://www.nrel.gov/analysis/sam/

[15]HOMER – The Optimization Model for Distributed Power, NREL, http://www.nrel. gov/homer/

[16]PVDesignPro, Maui Solar Software, http://www.mauisolarsoftware.com/

[17]Christensen et al. (2005). *BEOpt, Software for Identifying Optimal Building Designs on the Path to Zero Net Energy*, Presented at the ISES Solar World Congress, Orlando, FL, Aug 6-12.

[18]Energy-10 – Software tool for Architects, Builders and Engineers, NREL, http://www.nrel.gov/buildings/energy10.html

In: Solar America: How, What and When?
Editor: Nash M. Perales

ISBN: 978-1-60741-333-2

Chapter 4

ELECTRICITY GENERATION: HEARING BEFORE THE COMMITTEE ON ENERGY AND NATURAL RESOURCES, UNITED STATES SENATE ONE HUNDRED TENTH CONGRESS SECOND SESSION TO CONSIDER THE VALUE AND EXAMINE THE PROGRESS OF ELECTRICITY GENERATION FROM CONCENTRATING SOLAR POWER

U.S. Government Printing Office

The committee met, pursuant to notice, at 10:09 a.m. at the International Programs Building, Sandia Science & Technology Park, 10600 Research Road SE, Albuquerque, New Mexico, Hon. Jeff Bingaman, chairman, presiding.

OPENING STATEMENT OF HON. JEFF BINGAMAN, U.S. SENATOR FROM NEW MEXICO

The Chairman. OK. Why don't we go ahead and get started? This is a hearing of the Senate Energy and Natural Resources Committee that I am privileged to chair; Senator Domenici is the ranking member and Senator Sanders is a loyal member of our committee as well.

Let me just give a short statement and then call on Pete to make his statement, and Senator Sanders to make his. Then we will jump into the witnesses.

This is a hearing on a particularly important issue at this point in our history. I think a lot of the focus related to energy nationally is on the price of oil, and that is understandable. We are all worried about that, obviously, and we have tried to do some things with regard to that that we are still trying to add to in Washington.

Today we are focused on the issue of generation of electricity. For years, in fact, since 1978, the Federal Government has been trying to encourage the use of renewable resources for electricity generation. The frank truth is we haven't made much progress. Currently, about 3 percent of our electricity comes from non-hydro power renewables.

I think recently wind generation has been growing at a rapid rate, and over the last 2 years, it has been the fastest-growing source of new generation, which is encouraging. But that still leaves it with a very small share of our generation mix. Other technologies, renewable technologies have lagged behind.

Concentrating solar power, which is the subject of this hearing, has been pointed to as the next inheritor of the mantle that wind now wears. The primary reason that wind power has outdone other renewables is that it is cheaper than other technologies. It is about 8 to 9 cents per kilowatt hour, which is about the same as natural gas. Many in the industry believe that concentrating solar power could reach that price range within 5 to 10 years simply from the economies of scale that could come from installing a great deal of this capacity.

CSP, or concentrating solar power, has some real advantages over wind. The biggest is that it is less intermittent. In parts of the country where the solar resource is best, such as New Mexico, it is much more predictable than wind. It also may allow for easier storage of energy since it can use heat storage and does not depend on further development of battery technologies. That makes it more responsive to the actual needs of the users of electricity.

There are many reasons for us to pursue the goal of greater deployment of concentrating solar power as well as other renewable technologies. The Federal Government obviously has a significant role to play here. One issue that I know is on the minds of many here is the tax credits that we need to be extending in Washington, and I know there is going to be a continued push to get that done in the next month or two.

I believe, myself, that we should also enact a renewable electricity standard that would require utilities to acquire a set percentage of their electricity from renewables, and of course, we have that in New Mexico at the State level.

Another issue confronting all renewable generators is the inadequacy of the transmission system to carry the electricity that could be produced. We had a hearing on that issue a couple of weeks ago in Washington, and I hope that that hearing leads to some productive solutions on that score.

I wanted particularly to congratulate the utilities—Public Service Company of New Mexico, El Paso Electric, Xcel Energy, and Tri-State—for their announcement this week of a request for proposal for concentrating solar power generators to supply citizens here in New Mexico.

When we get to the witnesses, we are going to call on Greg Nelson from PNM to give us an overview of that proposed project and the reasons why they have chosen this technology to pursue. We have a panel of very distinguished witnesses today, and I look forward very much to their testimony.

Let me now call on Senator Domenici for his comments, and then on Senator Sanders.

Statement of Hon. Pete V. Domenici, U.S. Senator from New Mexico

Senator Domenici. Thank you very much, Mr. Chairman.

Might I first say how much we appreciate Senator Sanders being with us? Since you didn't welcome him, don't mind if I do.

[Laughter.]

The Chairman. You go right ahead, and I am well following you.

Senator Domenici. I wanted to tell him how different things are here than in your State. This is a much, much different State. But we know that you love what you have, and we want you to come here and learn about ours so you can love our State. Then all these things that Jeff needs in years to come when I won't be there, you will be voting "aye" every time.

Senator Sanders. Every time——

[Laughter.]

Senator Domenici. If you don't, he will remind you of the pleasantries of today and of the cordiality of New Mexicans. Let me say to everybody here, you know, I don't know why—no reason at all—but he and I have become friends.

[Laughter.]

Senator Domenici. We actually go out of our way to shake hands on the floor of the Senate, and that is kind of nice. We smile at each other and almost act like we are long-lost friends. That couldn't be because, you know, I am so much older than him, we couldn't be long-lost friends. But I do want to say that I not only appreciate you being here, but I very much appreciate the hard work you do. Whether I agree with you ideologically on whatever it is, I am a fan of your hard work.

Let me say I have a prepared statement that this wonderful staffof mine has prepared. But I think Senator Bingaman covered it. So if you don't mind, I am just going to put it in the record with your permission, Mr. Chairman.

The Chairman. Very good. Included.

[The prepared statement of Senator Domenici follows:]

Prepared Statement of Hon. Pete V. Domenici, U.S. Senator from New Mexico

Thank you, Mr. Chairman, for calling this hearing on an issue of enormous potential—Concentrating Solar Power. I am pleased to be back here at Sandia today and I'd like to welcome Senator Sanders from Vermont to our great state.

According to the Energy Department, "more energy from sunlight strikes the Earth in one hour than all the energy consumed by human activity on the planet in one year." I find that remarkable—just one hour of sunlight contains more energy than is used worldwide for an entire year. New Mexico is truly blessed to have this abundant, carbon-free source of energy—if we can harness it.

There is no question that America needs to become less reliant on foreign sources of energy. Over the last several weeks, I have been telling Americans that we are on the verge of economic destruction if we do not address the high price of gasoline.

I believe that we need to increase American production of oil by opening up new areas for exploration as soon as possible. But this oil production in the near term will provide us with a bridge to a clean energy future. On the other side of that bridge is an economy driven by clean energy technology like solar panels.

The good news it that our government has already invested literally billions of dollars for research and development of renewable technologies like wind, biomass and solar. In fact, for solar alone, we've spent more than $4 billion. And there are billions more to come for all forms of alternative energy.

One of the most important issues facing the solar industry right now is that tax credits that we passed in a bipartisan manner to help spur development are set to expire. It is of critical importance that they be extended—as soon as possible.

Congress has passed tax credits for renewable energy six times since 1992. Unfortunately, the Majority in the House of Representatives has decided to change the way we pass these credits by requiring that they be "offset" by tax increases on other industries.

Requiring "offsets" to provide tax credits for renewable energy makes little sense, particularly in light of a new study released by General Electric that shows that the renewable energy tax credits already pay for themselves. By spurring the growth of a new, clean technology industry, the renewable energy tax credits provide the government revenues from the projects' income, vendors' profits, and workers' wages.

In addition, Congress has never before required offsets for renewable energy tax credits. By setting a precedent that from now on, offsets will be required, the Majority is putting the renewable industry in the difficult position of having to fight for tax increases every single year. I believe that this is a bad precedent to set.

The Albuquerque Journal echoed this sentiment in a recent Op-Ed calling on Congress to move forward on a bipartisan bill. The Senate has already done so by approving the Cantwell/Ensign energy tax package by an overwhelming vote of 88-8. It is my hope that the House Majority will drop their objections to passing a clean extension of renewable energy tax credits and pass this bipartisan package as soon as possible.

Thank you, Chairman Bingaman, for convening this morning's hearing. Our impressive panel of witnesses are involved in all aspects of the solar industry—from performing the R&D work, to manufacturing plant receivers, to building the CSP plants, and finally, to purchasing the resulting electricity. I thank you all for being here today and I look forward to your testimony.

Senator Domenici. I just want to say thanks to all you witnesses who are lined up here. I know you have something to contribute, and we want to hear it.

I, myself, in preparing for this hearing, was a little bit aghast by the Spanish company. In their notebook, they had prepared a trend line of how long it takes from starting a plant to building one. I was quite amazed, and I hope you will tell us all about that.

I hope that isn't the national standard. I read your notes. It takes 6 to 8 years from the time you get ready to do one of these until you break ground. That is almost as long as a nuclear power plant, and I don't think they are quite as complicated. But perhaps I am missing something.

The other point I wanted to make is that it doesn't sound possible, but, you know, the Sandia National Laboratory was working on this kind of solar energy early in my Senate career. I mean, it was like 30 years ago we used to go out and look at the troughs, which is one kind of concentrated solar, and they would show us what the shortcomings were. Then they would solve those, and there would be another one.

But I continue to be amazed at not you, Sandia, but just at America, how long it takes to go from the research to the development to the actual application. Let me just say I don't think we can afford 10, 15 years on big energy projects that substitute even directly or indirectly for crude oil.

I don't think we can sit by and let those projects take 8, 10, 15 years to reach fruition because I believe the dependence on crude oil is real, and we can't get out from under it, no matter how hard we try, for a long, long time. That is a destructive thing. It is destroying our economy. This huge amount of money we send overseas is unbearable for the great country called America.

We won't stand to spend half a trillion a year for 10 years while we wait around for some substitute. Can't happen. We can't let it happen. Even your great project won't be around. But as it is, we aren't going to have the resources hanging on for so long. So I am very concerned about how long it takes for alternative energy development.

Last, but not least, I do know, without the testimony, that we must have a tax credit, the 30 percent that you all know we need, for the kind of solar we are talking about. I know some people blame us Republicans, and we Republicans blame them.

[Laughter.]

Senator Domenici. What we really have to do is we have to decide who is going to give. I don't know who is going to give yet and how long it will take. But we act one way based on what we think is a good rationale, and the Democrats—because of the House, Democrats in the Senate act another way, and we don't get anywhere.

But I will tell you, I support it. So don't put me in an "anti" position. I support it wholeheartedly. I don't know how to get around the problem unless Senator Bingaman and I decided that we might come up with a bipartisan substitute, and maybe we will. I don't know. We might try when we get back. If we did, it would go. It would pass. But we haven't been able to do that yet.

So please understand you can blame whomever you blame, but I have a good answer for why I can't vote for the way it is being done, and he has a great answer for why they are voting the way they are. We both want it done, and we voted 88 to 8 at the last vote for you. So you know it is bipartisan.

With that, thanks for calling the hearing, Senator Bingaman.

The Chairman. Thank you very much.

Let me just say a few words of welcome and introduction.

[Laughter.]

The Chairman. This hearing really is being held at the urging of Senator Sanders. He has been urging me now for several months to have a hearing on concentrating solar power. I suggested we do it out here in Albuquerque, which he was very agreeable to, but that somewhat explains his presence here. He is a great advocate for all renewable energy sources. In particular, this very much, I think, as his own statement will indicate, he is very much aware of the potential of this particular technology.

So, Senator Sanders, thanks for being here.

Statement of Hon. Bernard Sanders, U.S. Senator from Vermont

Senator Sanders. Senator Bingaman, thank you very much for welcoming me here. Senator Domenici, thank you very much for your kind words.

I just want to tell the people of New Mexico that you have two great Senators. At this particular pivotal moment in American history regarding energy, we are all fortunate to have Senator Bingaman in the chair.

I can also tell you that while Senator Domenici and I certainly have differences of opinion, there is nobody in the Senate, I think, who is more respected and better liked than Senator Domenici. I just want to thank you for your years of service not only to the people of the State, but to the entire country. Thank you, Pete.

This is an extraordinarily important hearing and an extraordinarily important moment in American history. I don't want to renew the debates that are taking place in Washington. I happen to believe global warming is real. It is of enormous consequences to this planet. If we do not get our act together, our children and grandchildren will be living on a planet with a significantly inferior quality of life.

What we are already witnessing is increased droughts, flooding, severe weather disturbances, the CIA and the intelligence agencies telling us that as water disappears, as food land disappears, there is going to be increased international conflict and war. I think we have no alternative but to move aggressively and boldly in addressing this international crisis. I want to see this great country once again be a leader in the world in going forward in terms of energy efficiency and sustainable energy as we do it.

So the bad news is we have a very serious problem. But there is good news out there, and the good news is that we know how to address this crisis.

The other very good news is that it is not just intellectual knowledge. We now have the technology to do that as a result of extraordinarily good scientific work, including the work done here at Sandia, and engineering work done all over our country. The tools that we need are moving forward aggressively, in my view, in energy efficiency, an area my State has done a very, very good job, and moved forward in such sustainable energies as solar, wind, geothermal, biomass.

I happen to believe that solar, especially concentrated solar, has extraordinary potential to produce huge amounts of electricity, and we are right in the middle—right here in New Mexico, Nevada—right in the middle of the area that has the capacity to do that. There are scientists and, I expect, people who will be testifying today who will tell us that within a reasonably short period right here in the Southwest, we could produce 15 to 20 percent of the electricity that we need in the country. In years to come, we can do even more than that. That is extraordinary.

Second of all, in terms of the cost of this electricity, what even the folks at Sandia are telling us that the more electricity we produce, the more we learn about this technology, the cost will go down. It will be competitive or maybe even be more competitive than other fuels that we are using today.

Third, as we clean up our environment, as we begin to reverse global warming through concentrated solar, through photovoltaic technology, through wind, through geothermal, through biomass, we are also going to create millions of good-paying jobs. Sometimes I get a little bit tired, the Senators and I hear folks are saying, well, the world is coming to an end, you know? It is going to be a great economic disaster.

As we move forward in new technologies, I believe absolutely the opposite. I believe there is enormous economic potential not just for the Southwest, but all over this country in creating good-paying jobs. I want to thank Senator Bingaman, Senator Domenici. On the energy committee, we have made some progress in the recent energy bill. But we have a long way to go. We do. But that was a good bill. It was a bipartisan effort, and we have made some progress.

Let me just, if I might, Mr. Chairman, voice one note of concern before we hear from our very distinguished panelists. Some of you may have heard that the Bureau of Land Management recently announced a moratorium on accepting new applications for concentrated solar plants. I think that that is very unfortunate. I will do everything that I can to rescind that edict.

Right now, there are approximately 150 applications that the BLM people are currently processing. To the best of my knowledge, not one application has yet been accepted. The reason that I hear, if you can believe it in the midst of this grave crisis of global warming, as I understand it, we have two people who are processing these claims. Two people.

So here we are, trying to reverse global warming, create millions of good-paying jobs, we have got a bottleneck with two people presumably working very hard. We have got about a $3 trillion budget. I think we can afford a few more people to process these claims, and we will be trying to do just that.

So I am excited about this hearing. In Washington, we all hear a lot about nuclear, and Senator Domenici and I may have some disagreements on that. We hear about carbon sequestration, and I have my doubts about that. We are not hearing enough about the potential of solar in general and concentrated solar in particular.

I think this is the most exciting, sustainable, energy concept that we have out there, huge potential. So I thank you very much, Mr. Chairman, for holding this hearing and look forward to hearing from our distinguished panelists.

The Chairman. Thank you very much.

I think, since this may take us a little time to get through all the witnesses, I am going to take off my jacket. I urge all the witnesses or panel members to do the same, so that we don't get too hot around here.

The beginning of this hearing is going to be Greg Nelson, who is with PNM. He is the person who is in charge of this recent proposal that has come out for a request for proposal. I asked if he would go first before the other witnesses and sort of frame the issue and indicate to us what has led PNM and the other utilities to this point of looking at this particular technology, this concentrating solar power technology and what they anticipate going forward with this.

So, Greg, thank you very much for being here on short notice. Please go ahead and give us an overview on this, and then we will call on each witness.

Our normal practice here in the committee is to give each witness 5 or 6 minutes to summarize the main points they think we need to know. We will include any statements you would like in the record in full, but I think that is the most useful thing. Then after we hear from all witnesses, we can ask some questions.

So, Greg, go ahead.

STATEMENT OF GREG NELSON, DIRECTOR, UTILITY SERVICES, PNM RESOURCES, ALBUQUERQUE, NM

Mr. Nelson. Thank you, Senator.

Senators, it is a pleasure and an honor for me to be here today to represent PNM, Public Service Company of New Mexico, to talk about our plans for solar power. We firmly believe that solar power is very important not only to our company, to our State, but also to the citizens of the United States.

With that, I would like to tell you a little bit about our RFP and the process that led us to issuing this RFP. Back mid-year of last year, we initiated a multi-utility study with the Electricity Power Research Institute as the project manager for that effort. EPRI engaged solar energy experts, engineering firms in the likes of Black & Veatch and Nexant, to support that effort.

We also had several Federal partners in that project that included the Department of Energy, Sandia National Labs, and NREL. We also engaged a State partner in terms of the New Mexico Energy, Minerals, and Natural Resources Department, and we also included environmental stakeholders in the process.

The multi-utility study, as I said, was headed up by EPRI with PNM as the initiating partner. Other utilities involved in that process were Xcel Energy and their affiliate SPS, El Paso Electric, Tri-State's Generating and Transmission, San Diego Gas and Electric, and Southern California Edison. The purpose of the study was to identify the most commercially viable solar technology for a central station solar plant located here in New Mexico.

After an exhaustive study, the conclusion was that based on the status of the technology, solar trough was the most applicable technology and a technology that was cost effective enough to move forward with a large project. Secondary issues related to the study included appropriate siting criteria, looking at water constraints, as well as transmission constraints.

Based on the results of that study, four utilities decided to move forward with a central station RFP. Those four utilities were PNM, SPS Xcel, El Paso Electric, and Tri-State. We have joined together because we believe, one, it is the right thing to do. The sum total of the four utilities represent the vast majority of the electric users here in the State. We believe that joining together for a large central station facility makes economic sense. It helps bring down the cost of the plant and, therefore, the cost to each of our customers.

We, this week, issued an RFP for a facility in the range of 211 gigawatt hours to 375 gigawatt hours. That range is equivalent to a facility without storage of approximately 110 megawatts up to 195 megawatts. With 6 hours of storage, we expect that would go from 65 megawatts to 120 megawatts.

Criteria that we have in place is that the facility must be located in New Mexico. We believe, according to NREL data, that we have the second-best solar resource in the Nation, and we want to take advantage of that for our customers and for the citizens of New Mexico. We issued the RFP, like I said, this Monday. We are giving respondents a couple of months to respond to it.

We hope to have a PPA, a power purchase agreement, negotiated and in place by the end of this year, and we hope to have a facility online by the end of 2011. That will not only help us meet our renewable energy RPS requirement, it will also help us bring clean energy to the citizens of New Mexico.

Senator Domenici. What will be ready by 2011?

Mr. Nelson. We hope by the end of 2011 that we will actually have a plant online and generating.

Senator Domenici. Delivering?

Mr. Nelson. Delivering power to our customers. Yes, sir. Fully commercial by the beginning of 2012.

The chairman. Very good. Did you have any more detail you want to give us at this point?

Mr. Nelson. I would be happy to open it up to questions. But we are putting it out on an energy purpose or on an energy basis, where we will be purchasing the energy off of that project along with the renewable energy certificates that go with it.

The Chairman. Are you requiring as part of the RFP—one question that occurred to me—that the 6 hours of storage be built into the unit so that there will be storage capacity?

Mr. Nelson. The RFP was formulated to be as flexible as possible to the vendors, looking for them to come up with creative ways to bring down the cost of energy as much as possible. We believe storage plays a large role in that. We have encouraged the RFP bidders to propose storage in there, to come up with a cost-effective level. But we do encourage them to look up to 6 hours of storage associated with the project.

The Chairman. Pete, did you have a question?

Senator Domenici. Yes. What is the nature of the property upon which the central plant would be located?

Mr. Nelson. We did not specify properties, although we have had a number of interested stakeholders offer up their property for the siting of this facility. Again, we wanted to leave that up to the bidders, again, to give them the flexibility to come in with the most cost-effective project they can.

Senator Domenici. Do you know in advance, having researched it, whether there will be right of way problems or right of way necessities that are going to have to be met?

Mr. Nelson. The right of way challenges we expect to face will vary depending on the location proposed. As I am sure you are aware, right of way and transmission constraints here in New Mexico are very significant. So there are certain parts of the State that are tougher to get the power back from than others. When I say "get the power back," I am talking to basically our load center, which here in New Mexico consists of the Albuquerque/Santa Fe area.

Senator Domenici. Mr. Chairman, I am sure that given time we will be inquiring, but not today. We don't have enough preparation, but I think it is good that you brought them. It makes the hearing much more relevant because we are right in the middle of an event of significance to us. Thank you for that.

The Chairman. Bernie, did you have some questions?

Senator Sanders. I did. Mr. Nelson, did I hear you say that you thought, if things go the way you wanted, you could break ground inline and be producing electricity in 2011?

Mr. Nelson. We hope to, if all goes well with our engaged partner—and we look at the ultimate person that we sign the PPA with as being a true partner in the development of renewable energy. We hope to break ground in the 2009, early 2010 timeframe and have the project online by end of 2011, worst case early 2012.

Senator Sanders. One of the exciting attributes of concentrated solar is the speed. In fact, I think Senator Domenici pointed out, purposely so, it is not a complicated technology. In fact, the generation aspect is traditional. It is what they do with coal and gas. The speed at which one can move these things, given the crisis that we face as a Nation, is one of the attributes of solar power. I think you have indicated that?

Mr. Nelson. Yes, sir. As you may be aware, a number of the large solar firms have done solar prospecting here in the State and identified sites, have supply chains set up. So we expect that construction should take, for a facility of this size, on the order of a year and a half timeframe. So that, with their site lined up, with their supply chain in place, we believe that the 2011 timeframe is a reasonable timeframe.

The Chairman. Thank you very much for being here and giving us that overview. Congratulations on moving ahead with this. I think it is a great project.

[The prepared statement of Mr. Nelson follows:]

Prepared Statement of Greg Nelson, Director, Utility Services, PNM Resources, Albuquerque, NM

Introduction

Good morning Chairman Bingaman, Senator Domenici, Senator Sanders, and distinguished Members of the Committee on Energy and Natural Resources. Thank you for inviting me here today. I am Greg Nelson, Director of Utility Services for PNM Resources.

PNM Resources is an energy holding company based in Albuquerque, N.M., with consolidated operating revenues of $2.4 billion. Our electric generation is primarily a mix of coal, nuclear, wind and natural gas. Through its utility and energy service subsidiaries, PNM Resources supplies electricity to 738,000 homes and businesses in New Mexico and Texas, natural gas to 470,000 customers in New Mexico, and electricity to numerous wholesale customers throughout the southwest. Its utility subsidiaries are PNM, TNMP and First Choice Power, a deregulated competitive retail electric provider in Texas. In November 2006, we announced a Joint Venture with Cascade Investments for the purpose of long-term investment in both wholesale and retail electricity sales, electricity generation and energy trading.

PNM Resources is committed to diversifying our generation. As Director of Utility Services, one of my main responsibilities is to oversee renewable generation, including wind, biomass, and most importantly for this hearing, solar, from inception to commercial viability for PNM Resources.

Solar

According to the National Renewable Energy Laboratory, New Mexico is one of the best solar resource capability in the nation. We firmly believe solar power is not only important to our company and to our state, but also to the US.

In 2007, we initiated a multi-utility study that included Xcel Energy, El Paso Electric Company, Tri-State Generating and Transmission Association, San Diego Gas and Electric, and Southern California Edison. Leading this study was the Electric Power Research Institute, who engaged solar experts and engineering firms in the likes of Nexant and Black and Veatch. We had several federal partners in the project, including the Department of Energy, Sandia National Laboratories, and the National Renewable Energy Laboratory. We also engaged the New Mexico Energy and Natural Resources department and environmental stakeholders including Western Resource Advocates and the Coalition for Clean and Affordable Energy.

The purpose of this study was to identify the most commercially viable solar technology for a central station solar plant located within New Mexico in the 2011 to 2012 timeframe. It was determined after an exhaustive study that, based on the status of current technology, solar thermal parabolic trough is the most applicable and cost efficient technology for a large project in this timeframe.

Based on these results four utilities, PNM, Xcel Energy through their affiliate Southwestern Public Service Company, El Paso Electric, and Tri-State Generating and Transmission Association moved forward with a central station parabolic trough Request For Proposal (RFP). The sum total of these four utilities represents the vast majority of the electric users in New Mexico. We believe that joining together for a large central station facility makes economic sense; thus bringing down the total cost of the plant for each utility, and subsequently lowering the cost of providing solar energy to each of our customers.

On June 30 we issued an RFP for a facility located within New Mexico in the range of 211 gigawatt hours to 375 gigawatt hours. That range is roughly equivalent to capacity of 110 to 195 megawatts for a facility without storage, and from 65 to 125 megawatts for a facility with six hours of thermal storage. We are expecting to enter into an energy only contract, i.e. no capacity payments, in which we also receive all renewable attributes.

We believe storage will play a large role in any future solar facility. Consequently, we have encouraged the RFP bidders to propose up to six hours of storage, but allow bid flexibility for cost effectiveness.

Our goal is to have a Power Purchase Agreement negotiated and in place by the end of 2008 and a solar facility commercially available by the end of 2011.

Conclusion

Thank you for your time and consideration. I would be pleased to answer any questions you might have and I look forward to being of service in any way I can to this Committee.

Footnote

The issue of time of use (TOU) rates was raised during the question and answer session. PNM does not currently have TOU rates for residential customers, whose usage typically drives utilities' peak loads. TOU rates are intended to pass real time pricing signals through to consumers, which allows them to make energy consumption decisions. Generation is typically dispatched on an economic basis meaning that cheaper sources of generation are initially utilized followed by increasingly more expensive sources. This translates to higher costs of energy during high usage periods. Consequently, TOU rates financially incent lower usage during these high cost times. Renewable energy resources that harness the sun's power typically align well with utilities' seasonal and daily load shapes, meaning that there is good correlation between solar availability and high use time periods. Solar plants produce the most energy during the summer months when energy demand is the highest, and less in the fall, winter and spring when energy demand is lower. With the implementation of TOU rates, the cost of solar generation will be compared with the higher cost generation that is online during high use times, thus making generation from solar resources more attractive.

The Chairman. Yes, Senator Domenici.

Senator Domenici. Could I just, for the record, make an observation about right of way? I think people might wonder if you and I worked awful hard to put together the Energy Policy Act, which includes a major section on right of way, how we get right of ways, when right of ways have run into a stone wall, so as to speak. You hear and I hear from some that we shouldn't have done that, that we shouldn't have provided a way to break the stalemate.

I want to say that those who think that Republicans ought to be against eminent domain when you run into a wall, I was heavily in favor of the right of way that we put in. You can attest to that, and I still am. If there are any of you up here who are going to be complaining that we shouldn't have done that, I want you to know that the more you complain, the more I am for changing it.

But I am for changing it to make it stronger because I think it is ridiculous for us to be holding up major projects because we have some ideological bent that States must control. We are in a crisis. We have got a big plant like this, and you have to cross State lines, and you have to find a right of way. I hope our section, which is really not very tough—we were pretty generous because we didn't want the bill to get killed on the floor of the Senate.

So, it is there. It is no question that the commission has the ultimate authority, and I am sure you are aware of that. You would probably vote for it again if we had it again, wouldn't you?

Senator Sanders. I think the point that you are making is everybody —you know, we can complain and moan and groan, but if we are going to go forward, we need to take bold action. That, by the way, is true with wind as well. In my State, I have to tell you, there is overall sentiment for wind. But people say, hey, we don't want to look at wind turbines.

Sorry. If we are serious about breaking our dependency on foreign oil, we are all going to have to do something maybe that we are not 100 percent for.

Senator Domenici [continuing]. It may save time.

The Chairman. OK. Let me just briefly introduce the rest of our witnesses here.

Charles Andraka is with Sandia National Labs. We appreciate you being here. He is going to give us an overview of this technology and how it compares with others.

Fred Morse is with Abengoa Solar, headquartered out of Spain, I believe, but very much sort of a leading provider in this area.

Mike Daly is right here with Mesa del Sol and is going to talk about their efforts to create solar power capability here at Mesa del Sol.

Alex Marker is with Schott Solar, which, of course, is in the process of building their production or manufacturing facility in Mesa del Sol, and we are very excited about that. He is going to talk about what they see as the prospects for concentrating solar.

Fong Wan is representing Pacific Gas and Electric, and they have been a leader in the use of solar energy technology. He is here from San Francisco. So we really appreciate him coming very much.

Mr. Andraka, why don't you start off and give us 5 or 6 minutes? Then we will just go down the line and hear from all of you, and then we will have questions.

STATEMENT OF CHARLES E. ANDRAKA, SANDIA NATIONAL LABORATORIES, ALBUQUERQUE, NM

Mr. Andraka. Thank you. Good morning, Senators.

My name is Charles Andraka. I am a distinguished member of technical staff at Sandia National Labs, and I have worked as an engineer in concentrating solar power, or CSP, for the last 23 years.

CSP, as you know, uses mirrors to concentrate sunlight to create an intense heat to drive a conventional engine or turbine. Why CSP? First of all, CSP is highly scalable. It is built with glass and steel, similar to our automotive and building industries. CSP makes grid-ready AC power because of the rotating machinery involved.

CSP is proven technology with over 400 megawatts of CSP deployed and operating in the Southwest United States at this point. CSP can incorporate storage. This allows smoothing of short-term transients as well as shifting the peak to match utility needs.

CSP can be hybridized to burn natural gas or other fossil fuels for firming capacity. CSP is efficient with a world record efficiency of 31.25 percent conversion of sunlight to grid-ready electricity in a commercial installation.

I want to talk about the potential for CSP or why CSP now, why the big interest? First, the Southwest U.S. has an incredible resource. We like to call this "the Saudi Arabia of solar energy." We have identified in easily reachable resources nearly 7 terawatts of generating capacity. That is about 7 times the current electric-generating capacity in the entire country.

The second area is the renewable portfolio standards, particularly in the Southwest United States. There is a large amount of solar needed to meet these standards, and the ramp-up rate can be met with CSP technology as part of a renewables portfolio. This would be combined with wind, PV, geothermal, and hydro.

Third is the current public interest in CSP. This is driven by the current energy situation as well as concerns about global warming.

Finally, the DOE laboratories' development over the last few decades has led to an unprecedented technology readiness for deployment of these plants. This has led to a current publicly announced deployment plans of 3 to 4 gigawatts of power, and that is equivalent to 6 to 8 new coal plants. Many more plants are in the initial planning stages.

I want to talk a little about the barriers to CSP deployment. The first is financial risk. As you know, these are large capital expenses rather than fuel costs distributed over the life of the plant. So there is financial risk up front, and there is technology uncertainty, particularly with the towers and dishes. The troughs have less of that uncertainty because of the deployed projects. The ramp-up to manufacturing at a high rate is needed to reduce the cost. So there is that chicken and egg syndrome.

The second barrier is taxation policy. The capital expenses on these plants tend to be taxed as property. California, as well as other States, waives the property tax on these systems and is leading the way in this area. The longstanding 10 percent ITC and other considerations level the taxation on an energy basis with conventional power generation. However, the 30 percent ITC would help overcome some of the initial financial risk.

The bottom line on the taxation policy is we need long-term financial policy stability for these plants to go forth. You have already mentioned the time it takes to permit. If the ITCs run out during that time, it is hard to get financing for these plants.

The third area has already been mentioned, and that is transmission capacity. This is the greatest consideration we are facing when siting new plants. We find we are not in the energy generation business, we are in the transmission business when we are trying to site these plants. The capacity of a current transmission grid is much smaller than the plants that are proposed and on the books at this point.

Senator Domenici. What is that?

Mr. Andraka. The capacity of the existing transmission grid is much smaller than the plants that are already on the books, these 3 to 4 gigawatts worth of plants.

Finally, the final barrier is the approval processes. The Federal, State, local, and utility processes are often cumbersome. There is nothing in place for the current onslaught of solar plants. The codes and standards need development. The local authorities are scrambling, trying to find a consistent code to apply to these plants because there is nothing out there like this.

I want to talk a little about the laboratory role. The laboratories play a critical role in facilitating the rollout of the concentrating solar power technologies into the marketplace. We are working hand in hand with industry right now. We also continue to play a critical role in the development of advanced systems with lower costs and higher performance.

We see a three-prong approach for laboratory involvement. A key effort right now is commercial development support, where we leverage the incredible laboratory experience to assist the commercial companies technically.

The second area is supply chain development, where we need to identify and exploit synergistic supply chains, such as the auto-motive industry. Along with supply chain development is supply chain development of personnel. We need to increase our involvement with universities and increase our internship programs.

A third area that the labs need to be involved in is revitalization of advanced development of CSP technologies. This leads to new technologies with step cost reductions, not just the manufacturing quantities. We need to do, as the laboratories of things industry doesn't even know they need yet. We have a significant track record over the last decade of identifying those areas, developing technologies, and those technologies are now being used by industry.

In summary, CSP has the potential to meet a large fraction of our energy needs in a portfolio with other leading renewables. The easy-to-reach resources in the Southwest are capable of 7 times the current generating capacity of the United States. CSP leverages existing U.S. manufacturing capabilities. Current market drivers have led to unprecedented interest in CSP.

Deployment acceleration requires improvements in taxation, regulatory, and approval processes and policies. The support of the national laboratories has been and will continue to be crucial to the success of commercial CSP projects. Continued industry support, supply chain development, advanced technology research, and stable policies will allow us, as a Nation, to take advantage of the tremendous energy resource identified in our own backyard.

Thank you.

[The prepared statement of Mr. Andraka follows:]

Prepared Statement of Charles E. Andraka, Sandia National Laboratories, Albuquerque, NM

Introduction

Concentrating Solar Power (CSP) describes a suite of solar technologies that use mirrors and thermodynamic processes to develop grid-ready electricity. Mirrors on tracking structures concentrate sunlight, producing high temperatures, which then drive conventional or novel engine cycles that in turn drive a generator to develop electricity. CSP technologies do not depend on strategic or high-tech materials, but rather are based fundamentally on glass and steel structures. The collected energy can be stored as thermal energy—an inherent advantage of CSP over photovoltaic solar and wind electrical generation. While CSP technologies are not as recognizable as photovoltaic power (PV) technologies, there are nearly 450 MW of CSP generation currently operating in California and Nevada, with additional planned deployments of over 3000 MW in the Southwest United States.

CSP has the potential to supply a large fraction of the energy needs of the United States, although prime generation sites exist primarily in the Southwest. Working in conjunction with other renewable resources established in other parts of the country, and with improvements to the grid infrastructure, the future of CSP in the nation's energy portfolio is indeed bright. The current cost of electricity generation by CSP trough plants is about $0.16/kWh. Other CSP technologies may produce lower cost electricity due to higher system efficiencies. With further technology development and increased deployment, the cost of CSP-generated electricity projected in several studies to reach $0.06/kWh. In addition, the high-temperature capabilities of CSP make possible highly efficient chemical processes that can lead to solar fuels production.

The DOE national laboratories, specifically Sandia National Laboratories and the National Renewable Energy Laboratory (NREL), have played a crucial role in existing CSP deployments, and we continue to work closely with industry to optimize and improve the designs and plans for upcoming deployments. The historical and ongoing technical achievements at the laboratories have been and will continue to be a cornerstone of successful cost reduction, performance enhancement, and deployment success. Key laboratory-developed technologies are deployed to the field by industry. The test capabilities at Sandia are unmatched worldwide, and provide a great resource to industry partners.

CSP Description

CSP converts the sun's energy into heat and then uses that heat to power an engine-generator unit. The sunlight is concentrated with mirrors—similar to concentration by a magnifying glass. The resulting heat is intense enough to create steam to drive a conventional turbine or to heat a working fluid in a smaller engine, similar to burning gasoline in an automotive engine. CSP technologies are large-scale, providing utility-scale generation of power, with near-term planned plant sizes ranging from 100 to 1000 MW. (A typical coal or nuclear plant may be 500-2000 MW.) CSP consists of three basic technologies: (1) parabolic troughs, (2) power towers, and (3) dish-engine systems. Each of these technologies uses a parabolic array of mirrors, on different scales, to create intense heat.

CSP is already being deployed, with 384 MW of capacity in nine plants in California and a new 64 MW plant in Nevada. Combined, these plants represent more than 140 plant-years of commercial operation. The national laboratories have continued to develop the CSP technology and have also helped improve the deployed plants. In 1998, the nine plants in California increased their rated capacity from 354 MW to 384, in part because of performance and operations and maintenance improvements pioneered by the laboratories. Over the last two decades, new deployments have been limited by the relatively low cost of electricity generation by natural gas. The recent dramatic increases in fuel cost, coupled with the Renewable Portfolio Standards (RPSs) in some states, have driven renewed interest in CSP deployment. The addition of the 30% Investment Tax Credit (ITC), as opposed to the 10% level, offsets some of the financial risks inherent in initial scaled-up deployments.

A key advantage of CSP is dispatchability (that is, the ability of a generating unit to increase or decrease generation, or to be brought on line or shut down at the request of a utility's system operator). Because the energy conversion process is a thermal action, the solar input can be supplemented in two ways. The first is through thermal storage, in which a working fluid is stored hot and then used when needed to drive the turbine. This process is very efficient, with over 98% recovery. The second is that systems can be "hybridized".where an alternate fuel such as natural gas can be burned to supplement the solar collection. This method is not as desirable as storage, but it does present an option that photovoltaic and wind energy sources do not provide.

A second advantage of CSP is the inherent "low tech" of the materials involved. The collection structures are typically steel or aluminum, with glass reflector surfaces. The resulting structures have been likened to "a funny-looking car." Indeed, several industry partners who work with Sandia have already leveraged the manufacturing capabilities of the Detroit-area automotive companies, as well as other basic American manufacturing companies.

Technology Descriptions

Parabolic Troughs

The parabolic trough system is a line-focus mirror array, as opposed to a point focus system. At the focal line, a specialized tube carrying a working fluid (such as a thermal oil or a molten salt) is heated. The working fluid reaches temperatures in the range of 500□ C. The collected heat can then be stored or directly passed through a heat exchanger to generate steam for a conventional turbine.

This technology is the most widely deployed CSP approach, and existing deployments help in obtaining funding and approvals for new installations. The state-of- the-art systems are solar-only (no storage), with an annual efficiency in the 12-14% range and a peak efficiency of about 16-18%. Typical plants in the past were sized under 50 MW due to power purchase agreement limitations. The newest plant, in Nevada, is a 64 MW installation. Proposed plants for Arizona are as large as 280 MW with storage. The larger size plants bring down the cost of the electricity generated through economies of scale.

Current trough research includes thermal storage development and testing, higher temperatures (which leads to higher performance), and lower cost designs. Key laboratory optical modeling and systems development approaches are helping industry to reduce costs without reducing performance.

One key component of trough systems is the receiver tube, a glass and metal structure that includes some laboratory-developed sealing technology. The Schott Solar Company is planning to build a plant in Albuquerque to fabricate this critical component.

Dish Engine Systems

Dish-engine systems consist of a tracking dish that concentrates sunlight to a single point, and a heat engine at that point which converts the intense heat to electricity through a rotating shaft and generator. Current designs center on a Stirling cycle engine, which is similar in many respects to automotive engines. Dish systems currently range from 3 to 25 kW capacity each, although larger systems are envisioned by some companies. Most companies currently developing dish systems intend to deploy fields of dishes, with aggregated capacities up to 1000 MW (for example, 40,000 25 kW dishes in one field). This deployment approach is seen as key to cost reduction.

Because of the point focus at each dish, the dish system is capable of very high temperature operation, typically in the 800□ C range (glowing red to orange). These high temperatures lead to very high system efficiencies for conversion of sunlight to grid-ready electricity. The current world record solar conversion efficiency is 31.25%, held by the Stirling Energy Systems 25 kW Dish-Stirling system located at Sandia National Laboratories. The annual efficiency of such a system is in the range of 22-25%.

Stirling Energy Systems has announced two large power purchase agreements in California. The first is with Southern California Edison for the energy from a plant with 20,000 dishes producing 500 MW, with potential expansion to 850 MW. The second is with San Diego Gas and Electric for the energy from a 12,000-dish system producing 300 MW, with possible expansion to 36,000 dishes and 900 MW. With recent investment, the prognosis for successful deployment is very good.

Current efforts in dish-engine deployment center on cost reduction and large-scale manufacturing. The role of the national laboratories in this effort is in technology transfer and design support. In particular, as non-solar entities are engaged to provide manufactured parts and systems, Sandia's experience is leveraged to be sure that solar performance is not compromised. Additional development is centered on alternate engine advancement that could lead to lower operation and maintenance costs. The large number of dishes deployed in single locations help ramp up the production rates, which also leads to lower costs.

Power Towers

The power tower is also a point-focus technology that allows for higher temperatures than those in trough systems. In the tower system, a field of steerable mirrors reflects the sun's energy to a large point at the top of a tower, where a working fluid is heated and then either stored or directly used to drive a conventional turbine. A commercial power tower is likely to be sized in the range of 100 MW electrical output, although both smaller and larger plants have been proposed.

Tower systems can directly generate steam at the receiver location to drive a turbine. Such plants, on a 10 MW scale, have been demonstrated in Spain, where they have achieved annual efficiencies in the 12% range. A second approach is to heat a molten salt working fluid to a higher temperature, then store this hot salt until the generation of electricity is needed. A small-scale pilot plant, operated in the 1990s, demonstrated the feasibility of this approach. Larger molten salt plants are expected to lead to 18-20% annual efficiency. The higher temperatures of the tower systems make the possibility of thermal storage more economically feasible than with trough systems.

Although no US power tower plants are currently in production or deployment, several US companies have recently announced plans to pursue and develop various power tower technologies. Additional research and development will concentrate on cost reduction of the tracking mirror systems (development which is likely to support all the CSP technologies) and on the development of robust, efficient receiver assemblies.

Storage

Thermal storage of energy is unique to the CSP technologies, and it represents a significant advantage over other intermittent renewable technologies such as wind and photovoltaics. The large-scale storage of thermal energy is highly efficient, with over 98% recovery of stored energy. (Compare this to the battery storage of electricity, typically in the 60-70% range.) In addition, the storage containment equipment and fluids are quite cost effective compared to batteries, and they are more environmentally benign.

The thermal storage uncouples the collection and generation phases of the CSP cycle. Energy can be collected throughout the day, with actual generation of electricity deferred until needed (for example, evening peak periods). With enough storage, CSP technologies will be able to provide baseload (continuous, around-the-clock) power generation in the future. In the shorter term, storage can firm up capacity during peak parts of the day as well as shift the generation to better match the utility's needs. Some utilities (for example, Arizona Public Service) have indicated they will not consider solar technologies without storage, as their peak period extends well into the evening. Other utilities do not see the need for storage in the immediate future, but begin to see the need as renewables reach toward 20% of the regional generating capacity. The use of substantial storage will allow CSP to provide greater than 20% penetration in the electric generation arena.

Trough and tower technologies are well suited to molten salt storage, a technology demonstrated in the 1990s on the Solar 2 pilot plant in Barstow, California at a 10 MW electric generation level. The demonstrated systems used a nitrate salt (which is essentially fertilizer) to collect and store the heat. Sandia National Laboratories is presently examining salts with the potential for a lower melting point (reduces parasitic loads and losses) and a higher operating temperature (improves total system efficiency). Sandia is also testing components and materials for durability in long-term exposure to the salt working fluids.

Development Needs

The current public interest, high energy prices, and state renewable portfolio standards are driving unprecedented interest in CSP technologies. Deployment proposals and plans, as well as private investment in solar technologies, have grown exponentially over the past few years. More than 3000 MW of known Power Purchase Agreements (PPAs) are now on the books, with many more reported to be in progress. These deployments are investor-driven, so risk must be minimized to support return on investment. The national laboratories are continuing to play a key role in technology deployment and personnel training. The accumulated knowledge and experience in the laboratories is being leveraged through partnerships, Cooperative Research and Development Agreements (CRADAs), and other mechanisms. This leveraging helps the commercial sector deploy effective technologies and minimize the waste of capital investment. However, the laboratories also need to revive a research and development role that will develop next-generation systems with the potential to meet long-term cost targets.

Support and development needs lie in three key areas. First, continued technical support of near-term commercial deployments is needed to leverage the DOE investment in CSP development. Second, supply chain development is necessary to bring US industry capabilities to bear on this key strategic resource. Third, the laboratories must continue advanced development, leading the CSP technologies to more cost-effective solutions that bring us to mainstream power generation.

Industry technical support has been and continues to be the cornerstone of the laboratory involvement with CSP. Tools, methods, and technologies developed at the laboratories are directly responsible for the feasibility of the proposed deployments, as well as for ongoing improvements of operational systems in the field. The CSP personnel base at the laboratories has been very stable when compared to other missions of the laboratories, providing a continuity and experience base unmatched anywhere in the world. We have demonstrated an ability to provide significant value to industry partners during design, development, testing, and qualification phases of these technologies. However, there are limited "experts" in the solar field, so the rapid expansion of CSP firms has led to a severe shortage of engineers with solar experience. Working hand-in-hand with the laboratories has proven a viable method to add to the "solar expert" ranks. Sandia National Laboratories has also made use of its expertise in other areas of the laboratory, including manufacturing, failure analysis, materials research, Supervisory Control and Data Acquisition (SCADA) system and controls development, information security, and systems engineering.With the large planned deployments, this aspect of CSP development is reactive to industry needs.

Supply chain development provides for a transition of US manufacturing capabilities to these new technologies. The CSP technologies are presented to potential cross-cutting suppliers to develop a manufacturing resource for use across the CSP spectrum. This approach allows the leveraging of existing US nonsolar suppliers, particularly in the automotive sector, rather than reinventing the manufacturing wheel. This approach has proven successful in several areas for the Stirling Energy Systems team. The engine is being "productionized" by a Detroit engine production firm. Very significant enhancements have been proposed that will reduce the cost of the engine, increase reliability, and improve the performance potential. There are unique capabilities in American industry, developed for other sectors, which will impact all areas of the CSP designs. Supply chain development also includes development of solar engineers through development of university programs and curricula. This aspect of CSP development must be cooperative with industry to leverage both laboratory and industry experience.

The laboratories must revitalize a thrust in advanced development for CSP technologies. Rapid deployment and substantial private investment make the CSP industry partners focused on near-term sales and deployments. Thus the laboratories must continue to develop next-generation systems, components, and tools. Industry is neither able to take on the risk of advanced development nor the distraction it would inject into the deployment process. Although industry has proposed approaches that will initiate large deployments, laboratory technology breakthroughs will lead to cost reductions that will make CSP technologies cost-competitive with conventional fossil-fuel power generation. The laboratories need to focus on development of disruptive technologies that will impact the cost and performance of CSP systems. Increases in system performance (efficiency) will directly impact electricity generation costs because the majority of the cost in these systems is in the collection apparatus (steel and glass). The laboratories must be proactive in the development of advanced technologies.

The laboratories have often developed new approaches that industry did not anticipate. These approaches often become part of the baseline technology that industry is prepared to deploy. Sandia has developed closed-loop tracking sensors and algorithms that substantially reduced the assembly accuracy requirements of dish systems. Rather than "perfect" installations, the closed-loop sensors and algorithms allow the system to learn and adapt to any imperfections, resulting in a substantial reduction in installation costs. Sandia-developed mirror facets have a substantially higher accuracy than prior "commercial grade" facets, and for about the same price. This development has changed the entire design paradigm for point-focus systems, as the improved performance has a substantial effect on the cost of electricity generated. These improvements are now entering the parabolic trough arena as well: Sandia-developed heat pipe receivers demonstrated a 20% improvement in system performance on one Dish-Stirling system. Further development is expected to bring this disruptive technology to the market. Systems models, tools, and development hardware have led to a better and more realistic understanding of system performance and costs. Spin-off technology and algorithms from Sandia's Advanced Dish Development System (ADDS) are being incorporated into the near-commercial products of Stirling Energy Systems, Infinia Corporation, and Eurodish. Sandia's new "TOP" (Theoretical Overlay Photographic) alignment system for troughs has demonstrated the benefit of optical alignment of existing trough plants, and it provides a tool to economically perform the alignment.

The high temperatures possible with the point focus systems (dishes and towers) make possible high-temperature chemical processes for the development of transportation fuels. Several processes have been proposed and are under development for splitting water using high-temperature processes, creating a reliable and cost-effective stream of hydrogen. Similar processes can be used to split CO_2 into CO and O_2. The CO can then be easily combined with hydrogen to create liquid fuels, which can then be distributed using the existing fuels infrastructure. The CO_2 could be supplied from sequestration at coal plants or, in the long run, through atmospheric scrubbing.

CSP Market Potential

CSP technologies are enjoying unprecedented interest and development, both in the US and worldwide. This interest is driven by a variety of factors creating something of a "perfect storm." Respected US and international companies are entering the CSP field, and significant private investment is flowing into CSP. In the US, there is significant solar resource in the Southwest states, primarily in areas with otherwise undesirable land.

Market Drivers

A variety of drivers have led to the current unprecedented commercial interest in CSP. The first is the Renewable Portfolio Standards (RPSs), primarily in the Southwest states, that mandate certain significant percentages of electricity generation must come from renewables. Although wind power has made significant deployments driven by the RPSs, utilities particularly like solar because of the match of the generation profile to the load profile. Therefore, as renewables have started to provide a notable fraction of the energy in some regions, the utilities have desired to balance wind generation with solar generation.

The second driver is the rapid and recent increases in fuel costs for conventional power generation. This factor is particularly applicable to natural gas plants, which were installed as "peakers" when natural gas was abundant and cheap just a few years ago. Currently the costs of CSP generation are very competitive with peak natural gas generation, even at relatively small deployment levels.

Third, the cost of all energy, especially gasoline, has driven public sentiment and support for solar energy. Not only is solar seen as a stable, US-grown energy source, but it is also "green," satisfying additional public sentiment concerning global warming and greenhouse gasses. The extraordinary public interest is demonstrated to me each day as I field numerous calls from the media and private citizens.

Finally, the investment by DOE and private industry over the last 20-30 years has provided a level of technology readiness suitable for significant investment in large deployments. Although technical and financial risk is still apparent, the technical risk has been reduced through the laboratory and cooperative projects, demonstrations, and technology development. Modern design, manufacturing, and analysis tools applied to CSP allow rapid movement from concept to feasible hardware while reducing costs and risk.

Solar Resource

Presently, CSP technologies require approximately 6 acres per MW of installed capacity, compared to non-tracking PV at nearly twice that requirement. This translates to a 500 MW plant using about 5 square miles of desert land. CSP technologies require "direct normal insolation," which is a measure of the brightness of the light coming directly from the sun, rather then reflected off clouds and sky. Therefore, CSP technologies work best in clear, dry environments like the Southwest United States. Figure 1* shows the tremendous resource available in the southwest states of New Mexico, Arizona, California, and Nevada, with some areas in Colorado. Obviously not all of this land is available for CSP deployments.

An NREL study filtered this data to exclude land already in use, environmentally and culturally sensitive land, and land with significant slopes. The remaining lands were only considered when contiguous areas were greater than 10 km^2 (or 4 mi^2) and a solar resource over 6.75 $kwh/m^2/day$. Figure 2 shows the filtered data. If the minimum direct normal considered is 6.0 $kWh/m^2/day$, a still very good resource, considerable additional land becomes available, particularly in the State of Utah.

Although the vast majority of prime land has been filtered out, there are still more than 53,000 mi^2 of land available for CSP projects. Table 1 shows a breakdown of potential land, filtered as noted, on a state-by-state basis. This analysis shows an available resource that is 7 times larger than the total nameplate generating capacity of the US electric grid. These data and maps are available from the Renewable Resources Data Center at NREL.

CSP Cost

Renewable resources are best compared on the basis of "Levelized Energy Cost" (LEC). This is the present value of the total cost of building and operating a generating plant over its entire economic life, which is then spread across all the energy generated during the life of the plant, resulting in an average cost per kWh of energy produced in present-day dollars. CSP plants do not have ongoing fuel costs, which represent a significant fraction of the LEC of electricity from conventional fueled plants. However, the CSP plants are highly capital intensive, essentially buying 20-30 years of fuel up front in the form of collection equipment. Therefore, the LEC of CSP energy is highly dependent on financing and tax structures, as well as the rate of production of the equipment being fielded. The value of CSP is impacted by the cost, but also by environmental considerations that may or may not have a financial value, such as carbon footprint. Policies in this area can impact the value of CSP substantially. The cost of conventional generation is influenced by the current high cost of fuels, which also helps the relative value of CSP.

Table 1. CSP generating capacity by state on filtered land.

State	Land Area (miles^2)	Solar Capacity (MW)	Solar Generation Capacity (GWh)
AZ	19,279	2,467,663	5,836,517
CA	6,853	877,204	2,074,763
CO	2,124	271,903	643,105
NV	5,589	715,438	1,692,154
NM	15,156	1,939,970	4,588,417
TX	1,162	148,729	351,774
UT	3,564	456,147	1,078,879
Total	**53727**	**6,877,054**	**16,265,609**

Figure 3 shows the anticipated LEC reduction for CSP projects compared to the cumulative deployment of CSP projects. This projection is taken from the Western Governors' Association (WGA) Solar Task Force Summary Report of January 2006.[1] This model uses a trough plant with 6 hours of thermal storage as a surrogate for all CSP technologies, and includes continuation of the ITC. Significant reductions in cost are expected through manufacturing improvements resulting from the sheer volume of deployed concentrators. However, significant supporting policy and financial assumptions are included as noted in the figure.

Private industry is leading the way on current deployments. As expected, the exact terms of the contracts with the utilities are closely held secrets, so it is difficult to obtain accurate current costs of CSP generation. However, the 2003 Sergeant and Lundy report is an excellent resource on the prognosis for cost reduction of trough and tower systems.[2]

This report is scheduled to be updated to include modern technology improvements and financial considerations, and extended to include the dish-engine systems.

The current LEC for trough plants is estimated at \$0.16/kWh. Industry experts have indicated that near-term deployments can be expected to produce an LEC in the \$0.12/kWh range (DOE semi-annual review conference, Austin TX, April 2008). Further technological developments and very large deployments are needed to reach the predicted \$0.06/kWh range. Several trough manufacturers have also indicated that the near-term deployments planned are highly dependent on a stable ITC policy.

The contract price for Stirling Energy Systems dish-system electricity is also a closely protected corporate secret. However, if one reviews current policy in California, it is clear that the base price for these near-term plants is likely at or below \$0.10/kWh.[3] The high efficiency of the dish-engine technologies makes \$0.06/kWh a feasible target. Again, the LEC is strongly impacted through large deployments leading to highly automated manufacturing. In addition, successful early deployments will lead to more favorable financing terms for later deployments, similar to the pattern seen in trough deployments.

No large tower projects in the US are far enough advanced to evaluate modern costs. However, the Sergeant and Lundy report indicates that towers can reach the range of \$0.055/kWh. Towers have the inherent advantage of simple storage combined with higher temperatures than troughs, leading to higher efficiency and therefore lower cost.

The current energy environment is encouraging substantial interest in CSP. More than 3000 MW of deployment has been announced in the Southwest United States, and additional large deployments are in the planning and exploratory stages. Table 2 lists the publicly announced deployments planed for the Southwest US. In the near term, we expect an ITC extension would facilitate these deployments and accelerate the cumulative deployment of CSP in the United States.

Table 2. Known planned deployments in the US as of Spring, 2008

Deployment	Size	Status
Trough/ORC in Arizona	1 MW	(APS Acciona, operating)
Trough electric project in Nevada	64 MW	(Nevada Power, Acciona, commissioned June 2007)
Dish-Stirling plant in Southern California	500 MW Option to 850 MW	(SCE, SES, Aug 2005)
Dish-Stirling plants in Southern California	300 MW Options to 900 MW	(SDG&E, SES, Sept 2005)
Trough plant	553 MW	(PG&E, Solel, July 2007)
Linear Fresnel Reflector	177 MW	(AUSRA, PG&E, Nov 2007)
Parabolic Trough with storage	280 MW	(Abengoa, APS, Feb 2008)
Arizona PS Consortium RFP	250 MW	(issued Dec 2007)
Parabolic Trough	250 MW	(FPL Energy, AFC filed)
Power Tower	900 MW	(BrightSource, PG&E, April 2008)
Other RFPs issued but not announced		
Total	**3,275-4,225**	

Figure 4 shows the expected impact of ITC extensions on the near-term deployment of CSP technologies, based on NREL projections published in the 2008-2012 Multi-Year Program Plan.[4] These plants take a number of years for design and development, permitting, and financing. Thus the importance of a stable, long-term taxation and credit policy cannot be stressed enough.

In summary, the cost of CSP is likely to be competitive with conventional generation processes. The cost reductions will come through a combination of technology improvement (performance improvement), design for manufacture (cost reduction through design), volume manufacturing (cost reduction through automation and stable factory orders), favorable financing (through investor confidence) and equitable taxation (recognition that the capital investment is comparable to fuel investment in a conventional plant). In the short term, the ITC will promote deployment to accomplish these cost reductions.

Market Barriers

While we enjoy an unprecedented renewal of interest in CSP, there remain several market barriers. If these barriers can be addressed, CSP deployments will accelerate more rapidly, moving the balance of our energy infrastructure toward a sustainable domestic resource.

The first barrier is financial risk. CSP plants are not consumer items; rather they are very large industrial complexes. The up-front cost is high, and it is paid back over long periods of successful operation. The lack of large deployments, particularly in dishes and towers, leads to uncertainty and therefore a higher cost for financing these projects. As plants are deployed, the financial risk is reduced, and the cost of financing is proportionally reduced. The current very high interest in troughs in part results from the ability to finance these projects based on the success of the over 400 MW in the Southwest United States. This is one area where the

ITC can significantly reduce the cost of the plant to offset the high cost of financing due to the perceived risk.

Similarly, taxation policies impact the financial feasibility of CSP plants. The high amount of capitalization results in a significant tax burden when compared to conventional-fuel power plants. A state policy to exempt these plants from property taxes will help level the playing field, making CSP competitive with conventional technologies.

As we work with companies in planning large deployments, we find that available transmission capacity is a much larger consideration than land cost. Despite high public interest in renewable energy, the public tends to be very opposed to new transmission capacity. A good example is the Sunrise Powerlink, proposed by San Diego Gas and Electric. Current transmission capacity can handle the introduction of the 300 MW Stirling Energy Systems dish-engine power plant in the Imperial Valley. However, the proposed extensions will need the Sunrise Powerlink, which is currently opposed by several activist groups. Beyond California, if we anticipate the Southwest United States supplying CSP-generated power to large portions of the country, substantial changes to the nation's electrical grid will need to be considered. Any new large-scale transmission lines will also face challenges in ensuring minimal environmental impact.

Many of the proposed plants are on federal government land, primarily BLM land. The permitting process for these lands, though necessary to protect various national interests, is a cumbersome and slow process. The shear size of these plants, several square miles each, presents unique environmental approval challenges that must be considered in detail. Streamlined permitting and approval processes for lands in the "CSP hotspot" could accelerate development and deployment.

BLM recently announced a two-year freeze on new solar projects on BLM land while they study environmental impacts.[5] This freeze forces the consideration of environmental impacts to be performed in series with other site considerations, rather than in parallel, effectively delaying new installations by another two years. A coordinated federal streamlined permitting process could significantly shorten the process leading to deployment, rather than the current patchwork approval process that adds significant delays.

Beyond the land-use permitting, site development and planning takes years in order to meet many state and local requirements. The technologies are substantially different than conventional technologies from a utility perspective. This is particularly true with systems that do not incorporate storage. Large intermittent sources have not been previously addressed by the utilities, so there is substantial uncertainty. There are no applicable codes and interconnect standards for such systems. All these significant technical and policy issues slow the approval process and add financial uncertainty to the project developer. We need sustainable energy policies, economic conditions, and permitting processes that motivate private investment in new technology deployment.

Conclusions

CSP has the potential to meet a very large fraction of our nation's energy needs, starting with grid-based electricity and expanding to transportation fuels production. The resource available in the Southwest United States on easily useable land is nearly 7 TW, or 7 times the current electrical generation capacity of the US. Cost-effective and efficient storage sets CSP technologies apart from key intermittent renewables of photovoltaic solar and wind. This is especially important as intermittent renewables begin to generate significant fractions of our national energy supply.

CSP leverages existing US manufacturing capabilities. The fundamental building blocks of CSP are glass and steel, materials common to the automotive and building industries.

Current market drivers—including global climate change, high fuel prices, and technology readiness—have led to unprecedented interest in CSP technologies. A number of US and International companies are poised to deploy large CSP plants in the Southwest United States.

Significant deployment acceleration requires policy improvements, including a stable taxation and regulatory environment and streamlined land use and interconnect approvals and policies.

The support of the national laboratories has been crucial to the technical success of CSP projects, and the laboratories' role will not diminish with the advent of large deployments. The partnerships developed between the laboratories and industry have been extremely valuable in the feasibility and success of new CSP deployments. Although the technical support of the deployments is critical, the laboratories also need to promote supply chain development leading to cost reduction, and they need to enhance long-term research and development of disruptive and advanced technologies that will dramatically impact the cost and performance of future plants.

Continued industry support, supply chain development, advanced technology research, and stable policies will allow us as a nation to take advantage of this tremendous energy resource identified in our own backyard.

The Chairman. Thank you very much. Appreciate that.

Mr. Morse, go ahead and give us your view from Abengoa.

Statement of Frederick H. Morse, Senior Advisor, U.S. Operations, Abengoa Solar, Inc.

Mr. Morse. Senators, thank you very much for inviting me to speak to you today on this very promising technology, CSP.

My company, Abengoa Solar, which has its U.S. headquarters in Denver, designs, builds, owns, and operates CSP plants around the world and would love to do the same in the United States.

As you mentioned, Americans are very concerned about their future. Energy prices are going up to unexpected levels, fears of recession, job losses, climate change issues. This CSP technology, as you have all acknowledged, is right in front of us, and it could be part of an energy, economic, and environmental solution. I think it should play a major role in our energy portfolio.

Charles mentioned what CSP is. I won't comment on that. But I will say it is bursting at the seams to come out on the Southwest utility market. Over a dozen companies are spending their own money to develop this technology without a Federal penny. Eight of them have signed contracts for over 4,500 megawatts, signed contracts. That could be instead of 4 to 6 coal plants or gas plants. These 4,500 megawatts could power over a million homes.

The utilities—and we just heard from PNM there, and we will hear from PG&E—they want to add CSP to their mix now, not tomorrow —now. The resource potential of the Southwest, your State and others, dwarfs the Pacific Northwest hydro resource. If it were oil, we would be racing to develop this. It can play a major role for the Southwest and the United States.

I can't resist. So I have to say there is one thing that stands in the way, and that is the immediate 8-year extension of the ITC. I will comment why. If that doesn't occur, New Mexico will lose. I think the facility that we heard about today will not get built. The Southwest will lose, and America definitely will lose. Without the 8-year extension, this industry will be stopped dead in its tracks.

With the extension of the ITC, I will state that there are no major barriers—there are problems—no major barriers for CSP to enter the utility market. Of course, there are other things the Federal Government can do—help with transmission, BLM we heard, siting issues, R&D support. Those are all in my written testimony.

The contracts signed to date, those 4,500 megawatts, are over $20 billion of investment. The investors are ready. They are ready to make the investment. But without the ITC, they cannot and they will not.

You asked about how long it takes to build a plant. I think building the plant, a year and a half is a good number. But permitting the plant could take another year or a year and a half. You have to win a contract with, say, PNM and negotiate that. You have to have a little pushing because the banks will say what happens on the day you start the plant up if the transformer blows and you need to buy another one? So that is the reason for the 8-year extension.

The economic benefits from unleashing CSP are impressive. Every dollar of tax credit will be multiplied many times by the investment to the $20 billion I mentioned—purchases from suppliers, wages from new jobs, and the flow through commerce associated. Those contracts on the books, none of which will happen—none of them will happen. My company has one, and we can't get it financed without the ITC—is 25,000 construction jobs. The manufacturing jobs, like the Schott plant and others, will all happen with the ITC.

I am going to close with an example. We negotiated a project with Arizona Public Service for a 280-megawatt plant with 6 hours of storage. This plant will deliver energy well past sunset while the peak is still there. It will power 70,000 homes. It is carbon free. It is on figure 6 in my written testimony. APS sees this as the first of many. If they can't get this plant built, they will go back to natural gas, I predict.

So America has an enormous domestic resource. It is ours. It is carbon free, and it is forever. We have to develop it because it creates jobs that cannot ever be exported, and the resource adds security of supply with diversity and so on. But we are risking that.

So my last word is CSP industry is ready with its money. Wall Street is ready with its money. The States are ready with their subsidies. All that is missing, I am sorry to say, is the Federal extension of the ITC, and I hope maybe you two can find a way to work out a deal. But if not, the industry is going to be severely hurt.

Thank you very much.

[The prepared statement of Mr. Morse follows:]

Prepared Statement of Frederick H. Morse, Senior Advisor, U.S. Operations, Abengoa Solar, Inc.

Senators Bingaman, Domenici and Sanders, thank you for inviting me to speak to you today about one of America's most promising renewable energy technologies — Concentrating Solar Power, or CSP. My company, Abengoa Solar, develops, builds, owns and operates CSP plants around the world and is also planning trying to do this in the United States.

Americans are deeply concerned over their future, with oil and natural gas prices rising to unexpected levels, with fears of a recession and loss of jobs, with reports that the concentration of carbon dioxide in the atmosphere continues to increase, threatening many adverse consequences. Importantly, the EPA has identified electricity generation from fossil fuels as the single largest source of domestic CO_2 emissions.

But there is an exciting clean energy technology right in front of us that can become part of the solution to America's energy, economic and environmental challenges. The solution is Concentrating Solar Power and it can and should be a part of our national energy portfolio going forward.

CSP refers to a family of technologies that convert the sun's thermal energy into steam to generate electricity with zero carbon emissions. Some CSP technologies concentrate the sun and convert it directly into electricity via an engine or photovoltaics located at the focal point. Figure 1** shows the major CSP technologies.

CSP is most cost effective at utility scale (hundreds of MW) and some CSP technologies can provide electricity, on demand, when it is needed, and some can even produce electricity well into the night to meet summer peak demand. Some CSP technologies are commercially available and have been working reliably for over 20 years in the Mojave Desert, where they have not failed to meet a single hour of peak demand since they came on line—with the help of favorable tax policies. Figure 2 shows the output from a portion of the 354 MW CSP Plant.

It can be seen that when combined with natural gas to firm the output, these plants had an on-peak capacity factor of over 100% every year of their operation. Most current CSP plants now firm their output using thermal storage to become "pure" solar plants. Utilities are familiar with CSP and wish to add it to their energy mix. Not tomorrow, but today.

CSP is a power system straining to burst onto the southwest utility scene. Well over a dozen companies are developing CSP plants using private—not Federal—funds, and eight have signed contracts with utilities which total over 4,500 MW—equivalent to 4 large coal or natural gas plants that will not have to be built. These 4,500 MW of CSP plants will be able to power over one million homes. Because CSP has attributes that utilities prefer (generates steam, comes in large sizes and is dispatchable), more utility contracts are certain as the cost of CSP declines relative to fossil-fuel generation.

The solar-rich Southwest can look forward to the day when a solar-powered plant, not natural gas-fired or coal-fired generation, will be a utility's first choice—irrespective of whether or not renewable energy mandates exist. Because the CSP resource potential in the Southwest exceeds the hydro potential of the Pacific Northwest, CSP can become a major driver of the economy of southwestern U.S. and play a major role in meeting the region's future energy needs and environmental targets.

As the rays from the sun enter the earth's atmosphere, a portion are scattered and absorbed by the moisture and particulates in the atmosphere while some reach the surface directly. The unscattered portion is called Direct Normal Insolation. Because CSP technologies can only use the direct radiation, it is essential to know where the level of that radiation is highest as that will be the best place to locate a CSP plant, assuming the site meets other requirements. The National Renewable Energy Laboratory has, over the past years, used satellite date to map the solar resource in the United States. Figure 3 shows the distribution of Direct Normal Insolation in the southwestern U.S. The darker the color, the higher the solar radiation and the better for locating a CSP plant.

NREL then used GIS methodologies to filter out places where a CSP plant could or should not be sited, such as cities, waterways, environmentally sensitive areas, and mountains and slopes greater than 1 %. The resulting map is shown in Figure 4.

While most of the areas with high direct normal solar radiation have been removed, what remains are the "sweet spots" for CSP in the United States. The remaining areas represent the upper limit because additional environmental restrictions may exist or be placed on their use for CSP plants. Figure 5 shows the area and potential for CSP generation on the areas shown in Figure 4.

It can be seen that there is ample land and potential in the southwest to provide as much electricity as is needed and desired. And the technology to use this resource, CSP, exists and is poised to enter the utility market in large amounts.

However, one federal legislative action is essential if this new wave of solar power plants is to happen: the immediate 8 year extension of the 30% federal Investment Tax Credit or ITC. Without an 8 year extension, this rapid growth of Concentrating Solar Power will not occur and New Mexico loses, the southwest loses and America loses. Even the 4,500 MW of signed contracts will be voided since their pricing is contingent on the long-term availability of the ITC. Without an eight year extension of the ITC now, the CSP industry will be stopped dead in its tracks. Only the U.S. Congress can extend the ITC, and this is something that has proved surprisingly difficult.

Why must the ITC be extended now? Because during the time since the power purchase contracts for the 4,500MWs of CSP plants have been signed, the price of steel has increased dramatically and similar increases have been noted in the other commodities used for CSP plants. Until the ITC is passed, financing is not possible and therefore it is not possible to purchase the components needed to build the CSP plant. Furthermore, the financial markets continue to be troubled, making debt financing more difficult and costly. The longer CSP projects have to wait, the more difficult it will be to adhere to the terms of existing contracts and to finance these projects.

Aside from the extension of the ITC, there are no major barriers facing CSP. If the eight-year extension is enacted, CSP will burst onto the utility market. Of course there are other things that the Federal government could assist the CSP industry with. For example, the BLM should adopt a friendlier land policy for CSP, a process that is has begun, thanks to an impetus provided by EPAct.2005 The BLM is now identifying large tracks of federal land that are well suited for siting CSP plants and will perform generic environmental studies. This process needs to move along rapidly and with input from the CSP industry, and needs to be adequately supported by the Congress. The Federal government should provide stronger leadership in transmission, as the lack of new transmission lines is affecting needed electricity growth in many parts of the Nation. This same inattention affects concentrating solar power as prime CSP areas lack connection to the growing load centers in the southwest. Furthermore, the long time it takes to build such lines will limit the rate and extent of the growth of CSP in the market, underscoring the need for greater action in this area. Finally, the CSP R&D program at DOE has been under-funded for many years and that needs to be corrected to support the innovation needed to help bring the cost of CSP electricity down.

The ITC could be thought of as a switch that if turned to ON and left ON for eight years, it will open the CSP market and trigger the building of CSP plants. The contracts signed to date with the utilities will require an investment of around $20 billion, all private sector dollars. And the investment community is ready to provide the needed debt and equity to finance CSP plants, but only if the ITC is extended for at least eight years.

The reason for the minimum of eight years is because of the long time it takes to complete the many essential steps to build a large power plant. These steps begin with finding and gaining control of a suitable site, obtaining regulatory approval on the power purchase agreement, completing the permitting process necessary to begin construction of the plant, building the plant and allowing some time for delays in any of these steps. This process is described in a presentation made at a briefing on 16 May 2008, at the request of Senator Sanders office.

CSP plants being bid today would be built in the 2012 and 2013 time period. If developers are to achieve cost reductions from 2 or 3 utility procurement cycles, only an eight year extension is effective.

The power point presentation titled "Why does it take so long to build a CSP plant" is attached to provide additional information on the steps and time needed to bring a CSP plant into commercial operation.***

The economic benefits from the unleashing of CSP are impressive. Every dollar of tax credit claimed by the ITC will be multiplied many times over in terms of the private capital investment, purchases from suppliers, wages for new jobs, and local and regional flow-through commerce. The job creation is significant. Approximately 25,000 construction jobs are associated with the 4,500 MW under contract. In addition, the building of new factories and assembly facilities for the main components will add more jobs and offer products for export to help our balance of trade. As natural gas is displaced by CSP, CO_2 emissions are reduced and, in time, its clean energy could be available for plug-in hybrid cars, thereby helping to reduce our dependence on foreign oil.

The following analyses provide additional details on the economic benefits form both CSP plants and related manufacturing:

- "Economic, Energy, and Environmental Benefits of Concentrating Solar Power in California." April 2006 by L. Stoddard, J. Abiecunas, and R. O'Connell Black & Veatch
- "New Mexico Concentrating Solar Plant Feasibility Study." February 9, 2005 Prepared for the New Mexico Energy, Minerals and Natural Resources Department by the University of New Mexico Bureau of Business and Economic Research and Black and Veatch. The study may be found at: http://www.emnrd.state.nm.us/ ECMD/ Renewable Energy/ documents/NMCSPdraft-final-rpt-02-05. pdf
- "The Potential Economic Impact of Constructing and Operating Solar Power Generation Facilities in Nevada: Draft Preliminary Report." July 8, 2003 by R. Keith Schwer and Mary Riddel of the Center for Business and Economic Research University of Nevada, Las Vegas

I want to close with a specific example of CSP's potential:

If the ITC is extended, Abengoa Solar will build the 280MW Solana CSP plant in Arizona that is now under contract. Under this contract, the energy will be sold to Arizona Public Service, powering over 70,000 homes with carbon-free energy. The schematic in Figure 6 below shows the Solana CSP plant. This plant will create 2,000 construction jobs and about 85 permanent jobs in a town with 68% of its population living below the poverty level. And this plant is likely to the first of many that APS will build to meet its growing demand for new electricity. However, Solana will not be built unless the ITC is extended soon. Instead, APS will likely turn to natural gas, adding to the climate change issues associated with fossil fuel generation. I attach for the record, a letter sent to Senator Cantwell, explaining why Solana can not be built unless the ITC is extended now.

American cannot afford to ignore one of its greatest domestic energy resources, especially if it is carbon free and can never be depleted. America cannot afford to ignore developing that energy resource, especially if it will create jobs that can not be exported. American cannot afford to ignore that resource, especially if it adds to security of supply and to the reliability of its energy system. But we risk doing just that.

CSP developers are investing their money to develop CSP projects, Wall Street is ready to provide debt and equity, the states have invested in CSP with their incentives —the missing and critical investment is that of the federal government via the ITC. Please extend it now.

Thank you for your attention. I would be pleased to answer your questions.

Attachment.—Letter to Senator Cantwell

APS,
ABENGOA SOLAR,
June 25, 2008.

Hon. MARIA CANTWELL,
511 Dirksen Senate Office Building, Washington, DC.

Dear Senator Cantwell, Thank you for your long standing support for the commercial solar power industry. Two months ago, you and Senator Ensign successfully amended the Senate's housing stimulus bill to include language that extended several alternative energy tax incentives, most notably I.R.C. Section 48's Investment Tax Credit for solar investment. The vote, 88-8, clearly demonstrated the support for incentives for alternative energy. Your strong leadership on this issue is greatly appreciated.

This letter is in response to your request for information from Abengoa Solar about the impact of the failure to enact an eight-year extension of the 30% investment tax credit (ITC) for solar property would have on the proposed Solana plant to be built near Gila Bend, Arizona. The Solana plant is a 280 MW Concentrating Solar Power (CSP) plant that is scheduled to be brought on-line in 2011. Abengoa Solar, Inc, a U.S. corporation, will own and operate this plant. It will require over $1 billion in capital investment, will create about 2,000 construction jobs and about 85 full time jobs to run the plant once it is built. In addition, if Solana were to be built, Abengoa Solar would also build an industrial mirror factory that would create some 150 jobs in the Southwest.

The output from this plant will be purchased by Arizona Public Service (APS) under a Purchase Power Agreement and will provide the new capacity needed to meet the state's growing demand and to respond to its requirements for clean energy generation. This plant will use thermal energy storage to allow the plant to continue operation for six hours after the sun has set, allowing APS to meet the summer cooling demand well into the evening hours.

Because the Solana plant is a partnership between APS and Abengoa Solar, both companies are responding to your request.

A long-term extension of the 30% solar ITC is needed to allow Abengoa Solar to provide the electricity at a price that APS is able to pay. If Congress fails to enact an 8-year extension of the ITC in the coming weeks, the simple answer is that banks and equity investors will be unable to provide financing for this plant and it will not be built. In the absence of a long-term extension of the ITC, APS would more than likely rely upon a natural gas fired plant to meet the demand that could otherwise be met with clean, solar power from Solana.

The delay and failure to pass the extension of the ITC has other consequences. Since the price for the electricity from the Solana Plant was fixed several months ago, the price of steel has increased by over 30%. As this plant will use as much steel as would be needed to build a second Golden Gate bridge, this also risks the economic viability of the plant. Further delay will only exacerbate this situation. To be very clear, if the ITC is not extended soon, the Solana project will not be built because the financing will not be there. The same fate is very likely for the almost 4,000 MW of other CSP plants with signed contracts, based on the availability of the 30% ITC. The total loss of investment is close to $20 billion and the loss in related jobs is well over 20,000.

Additionally, we need to be clear that the Solana plant is relatively far along in the development process and that many of the other proposed CSP plants that together represent our national hopes for utility-scale CSP are further back in the development process and will require the proposed eight years to qualify for the project financing.

We have followed each of the back and forth efforts by the House and the Senate to extend energy tax incentives. You know that the American public is watching this as the price of fossil fuel supplies continue to rise and they want to see a long-term national vision that will transition us towards sustainable clean energy. The failure of Congress to find a way to pass this extension is jeopardizing U.S. jobs and economic activity. We hope some compromise can be found that would allow this extension to pass very soon so Solana can be built and its benefits to the economy and environment can be realized.

Sincerely,
SANTIAGO SEAGE,
CEO, Abengoa Solar.
DON BRANDT,
CEO, Arizona Public Service.

The Chairman. Thank you very much.

Mr. Daly, go right ahead. Tell us Mesa del Sol's involvement in all of this.

STATEMENT OF MICHAEL DALY, MESA DEL SOL, ALBUQUERQUE, NM

Mr. Daly. Thank you, Mr. Chairman, Senators. It is an honor to be here today.

My name is Michael Daly. I am with Forest City, a national developer, and we are developing a project called Mesa del Sol. I am here really to speak on behalf of economic development primarily because I think we have got a good pulse on what is going on.

When we came to New Mexico about 3 or 4 years ago, our first job was to create high-paying jobs for New Mexicans. We wanted to build a sustainable community with some affordable housing and also concentrate on creating a community of continuing learning. We think the solar industry has created a unique opportunity here in New Mexico, which could be mirrored in other southwestern areas.

We started with a design. All of our buildings are LEED certified. We are doing Energy Star on the housing, and we are doing some unique water-harvesting ideas. But with the help of Sandia—and I look at my colleague Chuck, who taught us what concentrated solar was 3 years ago in his lab, and they have been a tremendous resource—we have gone further.

We are working with the labs on creating the first smart grid at Mesa del Sol, so houses can actually follow up or down their electricity depending upon time of day and peak generation needs. We are looking forward to having about a 40-house nanogrid at Mesa del Sol smart houses as a demonstration project. We have 38,000 houses we can leverage over time to roll on a very aggressive program.

We are also working with Bob Galvin and the labs on a perfect energy circuit at PNM and looking at transmission and how we run our switches to create a more efficient system. Now the base electric cost for Mesa del Sol is $180 million. It costs $280 million for the perfect system. It is not that we are going to spend the extra $100 million, but we can provide for the future, and PNM has been very helpful with that.

But probably most important, we have learned with the Sandia people and through the delegation as well as those in the State is that Mesa del Sol has an ideal site for a concentrating solar site, which we hope to respond to the upcoming PNM RFP.

That is actually controlled by DOE. It is a 1-by-5-mile buffer that we have been working with DOE and the base on, as adjacent to the base and part of the load center. So we will be responding to the concentrating solar RFP, hopefully with a development partner. We are talking to some today. We think it is terrific to have that model project here in Albuquerque. It works economically, and it is the best site for PNM.

But on the economic development front, what we did 3 years ago when we came here is we started to canvass solar companies. We went to trade shows. We went to sites, and we really searched out—and my colleague Dr. Marker and I had dinner and a lunch in Colorado 3.5 years ago talking about Schott, which was really the beginning of that deal.

The solar industry is a terrific industry for jobs. They are high- paying jobs. New Mexico is well suited for it. They have got the university. They have got the labs, and there is a natural propensity to do that type of research.

I will say it started with Advent Solar, which the delegates were very helpful with, which we appreciate, which is up producing solar modules itself in an 87,000 square foot facility here. That is 300 jobs.

I want to take a moment to note that the biggest problem is solar engineers, and we are working with a project at UNM to try and get some engineering programs. Forest City has currently endowed a chair in digital and film media. We are going to endow a chair and work with Schott to create education programs for solar engineers. So I think that cornerstone of educating engineers to create the solar is really important so that my colleagues, for all of us to get the students to do this technology.

The Schott Glass was literally the shot heard around the world. After the Schott Glass deal, our inquiries from solar companies went from approximately 1 a month to 10 to 15 a month. It is interesting in terms of where the economy is, our inquiries and our canvassing relative to actual industries that are expanding has really compressed in the past 2 months, but the solar business is going crazy.

Mirror companies, receivership companies, balance assistance, photovoltaic companies are all looking in the Southwest and looking to expand. But a lot of them are hesitant to make major commitments. I am pleased Schott has made the commitment to make a plant until the ITC has passed.

So we are looking forward to continuing the economic development. We are looking forward to creating our cost. The Sandia lab is important, and university education is important.

Fred has mentioned some economics, but I just wanted to give you some broad-brush things. About 0.1 percent of our electric is generated by solar nationally today.

Senator Domenici. One tenth of 1 percent?

Mr. Daly. One tenth of 1 percent. So it is a relatively small amount of our power.

Senator Domenici. That is both the kind that we are talking about today and the other kind?

Mr. Daly. PV.

Senator Domenici. All kinds of solar?

Mr. Daly. All kinds of solar. So it is a miniscule amount. To the extent that we went to 2 percent of our power, or approximately 6,600 megawatts, which is close to what is on the books right now, we would create more than 1,000 permanent jobs, 25,000 construction jobs.

The real rollout that I find important is the Schott factory could generate enough receiverships for that, or Schott's competitors, that is another 1,400 permanents jobs. For the mirror company that has got to come here to do the manufacturing, that is another 1,400 jobs.

You know, with a multiplier effect, that winds up being 40,000 jobs that could be created by just doing this 5,000 megawatts of electricity, which is a tremendous economic boost not just in New Mexico, but nationally. There are photovoltaic companies in Vermont, in Boston, in Massachusetts, all over the United States that are waiting for this to expand.

The total construction dollars to create this 5,000 megawatts is about $40 billion in construction, which really would just multiply through our economy. That is a why we should do it from an economic development point of view. We would certainly be proud to have at Mesa del Sol a 600-acre concentrating solar plant that is located adjacent to the base of the research facility. We think it would be a tremendous coup as the radiation values are there.

What do we need to do? Everyone has talked about the ITC credit. I would like to mention something, which is a unique opportunity for DOE to get involved. DOE currently consumes 15 megawatts of electricity a day, has a 15-megawatt load here. But when you go to DOE, they are hamstrung, from a policy point of view, that they can't enter into a long-term power purchase agreement. So they, themselves, cannot enter into and facilitate a renewable power contract for 30 years because they are hamstrung with the current procurement rules.

Secondarily, their proxy for what they can purchase that power for is today's fuel prices with no adjustments into the future. So I find it an opportunity as much as a detriment to have DOE take a leadership role in here and perhaps in the energy bill create ways that the Federal Government, DOE, can be like the utility company and agree to pay a premium.

While the power is maybe priced at 6 or 7 cents a kilowatt hour now, perhaps it would have to be 13 or 14 cents a kilowatt hour, which is detrimental for the military budget. Whereas DOE, they are the model of—as we were saying, I think it would be good if they took it on themselves. The load is right here, and there is a possibility of them participating as well, as well as other major military installations in this solar belt.

Education. I think creating centers of excellence and getting funding for specific photovoltaic programs at our universities, our resource institutions is terrific.

Ultimately, I have to say, as a developer, as an entrepreneur who wants to see things happen, there is an instance—as Fred mentioned, there is a problem getting permitting. The DOE controls one of the many sites in New Mexico to make one of these solar facilities a reality. It would be a shame if DOE was actually the impediment in creating a longer schedule versus being the facilitator to make one of these things going.

I do want to say that I appreciate the delegation's offices. I have visited them several times. We wouldn't be as far in both education and knowing what we have to do, and some of these things are already in the works, but I wanted to say these are things that can really facilitate solar jobs, which is our main mission.

Thank you.

[The prepared statement of Mr. Daly follows:]

Prepared Statement of Michael Daly, Mesa Del Sol, Albuquerque, NM

Mesa Del Sol Concentrated Solar Plant

- Site control of a 1200 acre flat site adjacent to Mesa del Sol.
- With a 800 megawatt generation projected deficit, PNM has authorized a feasibility study which will examine the feasibility of building a concentrated solar plant in New Mexico by 2011.
- PNM has established a joint venture with EPRI to investigate developing a plant.
- Last year New Mexico State Legislature put in place significant incentives which will subsidize the cost of producing electricity via a concentrated solar trough plant.
- Sandia Labs is the leading concentrated solar trough research institution in the United States and is located immediately adjacent to the Mesa del Sol site.
- Mesa del Sol is part of the largest load center in the state and transmission to the grid is readily available on site.
- Schott Glass, as a new employer in New Mexico, is a leader in receiver tube technology
- Mesa Del Sol has Among the best solar Resources in the United States.
- The Mesa Del Sol project would have minimal Environmental Impacts

- Kirkland Air Force Base could buy green power from the plant.

The Chairman. Thank you very much.

Dr. Marker, tell us your perspective from Schott Solar.

Statement of Alex Marker, Research Fellow, Schott North America, Inc., Elmsford, NY

Mr. Marker. Senators, thank you for the opportunity to speak with you today about concentrating solar power and the vital role CSP can play in securing America's energy independence, creating jobs right here in New Mexico, and how CSP can become an economic engine, driving sustainable growth.

In just 1 hour's time, the amount of energy that the Sun shines upon the Earth's surface exceeds the energy consumption of all mankind in an entire year. In the 5 minutes I will be speaking with you today, the Sun shining upon the United States alone contains enough energy to satisfy Americans' power demand for an entire month. Nowhere is that potential greater than right here in the desert Southwest and especially in New Mexico.

CSP is a proven technology. The SEGS plants in the Mojave Desert have been operating for more than 20 years providing 350 megawatts of power, generating clean electricity to hundreds of thousands of homes in California. Just last year, 6-megawatt Nevada Solar One facility went online producing clean electricity for Nevada.

There are plenty of high-value resources available right here in New Mexico for CSP generation. In fact, the entire State's electrical needs could be satisfied by the Sun. Energy can be even exported to other regions.

But all of this is at risk without a firm commitment from the Federal Government in the form of effective policy and long-term extension of the investment tax credit. The U.S. is a sleeping giant when it comes to solar energy. By extending the investment tax credit, this giant will awaken.

According to independent analysis, there are approximately 4,000 megawatts of CS power plants currently in planning or development stage. That is 4,000 megawatts that will most likely never be built without a long-term extension of the ITC. The industry needs the Federal Government's support to make CSP power generation competitive with that of traditional generation technology.

What does effective Federal policy translate to? For one thing, the increase in solar energy adoption means an increase in jobs. It is forecast that if the ITC is extended, 62,000 manufacturing and distribution jobs will be created in the solar industry directly as a result of increased adoption of renewable energy. Many of these jobs in the CSP arena.

On top of that, there will be an increased demand for electricians, plumbers, engineers, potentially thousands of new jobs created each year. This is job growth for Americans by Americans for an industry that will benefit America.

The company I represent, Schott Solar, is in the construction phase of a large manufacturing facility in Albuquerque. This plant will employ 1,500 people in the production of photovoltaics and receivers for the CSP power plants. Over the long term,

Schott's investment in New Mexico will reach $500 million, and the economic impact is forecast to exceed $1 billion.

This is just what one company is doing in one community. There are other companies undertaking similar large projects from Michigan to Oregon and many more that are ready to do so once a clear commitment from the U.S. Government is established in the form of a long-term investment tax credit.

If the renewable energy tax credit expires, the impact next year will be more than 100,000 jobs either lost or not created, according to Navigant Consulting. Additionally, there will be more than $20 billion worth of investments that won't be made, and there is no doubt that this money and those jobs won't go overseas.

Renewable energy is domestic energy. Domestic energy not only means jobs for Americans, but it means reducing our greenhouse emissions. It is something Americans want. According to a recent poll, 94 percent of Americans say that it is important for the United States to develop and use solar energy. Almost 80 percent feel that the Federal Government should make development of solar energy a major priority through actions such as extending the ITC.

We have an opportunity today to address the challenge of global warming while growing our economy. All we need to do is harness the power of the Sun, and to do that, we need your support. Distinguished members of the Senate Energy and Natural Resources Committee, we sincerely thank you for your time and your consideration in this important matter.

[The prepared statement of Mr. Marker follows:]

Prepared Statement of Alex Marker, Research Fellow, Schott North America, Inc., Elmsford, NY

Summary

- Concentrated Solar Power (CSP) represents a proven and reliable technology.
- Solar energy is relevant for almost every country in the world, especially the United States, where conversion of only 2.5% of the nation's usable area into solar farms would satisfy the entire nation's energy needs.
- Investment in solar will lead to the creation of hundreds of thousands of jobs (UC Berkeley).
- Energy produced from the sun by CSP benefits from stability in costs, as there are no commodity priced raw material requirements for fuel, only minimal (3 cent kW/h) operating costs.
- By 2050, solar power could end U.S. dependence on foreign oil and slash greenhouse gas emissions (Scientific American).
- With the necessary investments, energy produced by the sun could become cost competitive with fossil-fuel based technologies by 2020 (NREL).
- The United States has the opportunity today to address the challenge of global warming while creating jobs and growing the economy.

A Proven Resource with Almost Limitless Potential

In just one hour's time, the amount of energy that the sun shines upon the earth's surface exceeds the energy consumption of all of mankind in an entire year. In the time it takes you to read this document, the sun shining upon the US alone contains enough energy to satisfy America's power demands for several months. Energy from the sun is an integral part of a renewable energy portfolio. A portfolio that would strengthen our nation's economy, secure our energy independence, and provide clean energy to meet the ever increasing demand.

That potential is greatest in the desert southwest, and especially New Mexico.

The idea of harnessing the power of the sun is not new. Documents dating back to Archimedes have shown theories on how this can be accomplished. Yet it hasn't been until recently that major strides have been made on mass-producing solar technology, and not until the last few years that technological innovations have been made to dramatically reduce costs.

Concentrated Solar Power

Concentrated Solar Power (CSP) plants are utility-scale power plants that generally produce greater than 50 MW of power, enough to supply the energy needs of thousands of homes. In one variation of CSP, called parabolic trough, hundreds of trough-shaped parabolic mirrors are continuously adjusted to face the sun. These parabolic mirrors concentrate the sun's thermal energy onto receivers, located along the mirrors' focal points.

The concentrated solar radiation increases the temperature of the thermo-oil Heat Transfer Fluid (HTF), flowing through the receivers, to approximately 750°F. This super-hot fluid is then used to turn water into steam, which drives a turbine, generating electricity. The capacity of these power plants is well suited for utility-scale power generation as the plant's peak efficiency matches peak demand requirements placed on the grid.

Reliable and Proven Clean Electricity Generation

Over the decades, solar technologies have been reliably providing clean energy to tens of thousands of Americans. Photovoltaics have been in production for 50 years, and SEGS in the Mojave Desert, a CSP parabolic trough power plant, have been operating for more than 20 years, providing 350 mega watts of power per year. Just last year the Nevada Solar One facility went online producing 64 mega watts of clean power.

United States—A "Sleeping Giant"

The U.S. has at least 250,000 square miles of land in the Southwest alone that are suitable for constructing solar power plants, and that land receives more than 4,500 quadrillion British Thermal Units (BTU) of solar radiation a year. Converting only 2.5% of that radiation into electricity would match the nation's total energy consumption in 2006.

According to the American Solar Energy Society: "Generation from CSP technologies, especially those that can be augmented with thermal storage or hybridized with natural gas, is well matched with southwest load profiles, which tend to peak in the late afternoon and early evening."

"States with suitably high solar radiation for CSP plants include Arizona, California, Colorado, Nevada, New Mexico, Texas and Utah. Even if we consider only the high-value resources, nearly 7,000 GW of solar generation capacity exist in the U.S. Southwest." (Jan. 2007)

According to independent analysis, resource calculations show that just seven states in the U.S. Southwest could provide more than 7 million MW of solar generating capacity—roughly 10 times the total U.S. generating capacity from all sources today.

The following chart shows available resources in the desert Southwest (considering grade of less than 1 degree, and other necessary land conditions.

State	Available Resource	Land Area (mi^2)
New Mexico	1.6 TW	12,800
Arizona	742 GW	5,750
Nevada	619 GW	4,790
California	1.1 TW	9,157

Source: National Renewable Energy Laboratories (NREL). SunLab

The DESERTEC model, which has been developed to supply solar energy to Europe, provides a realistic model in exporting energy. In the Desertec model, energy would be generated in Northern Africa and Southern Spain and then shipped to Northern Europe. A similar model can be adapted and applied to the Southwest of the United States, where states like Nevada and New Mexico export solar energy to northern areas of the U.S. and Canada.

A key stumbling block in the US however, is in transmitting the energy produced in the Southwest to other regions. The need for a national "smart grid" is seen as essential in creating a network of energy produced by renewable energy. Even without the proliferation of renewables, many experts are in agreement that the nation must implement a "smart grid"—as evidenced by the California rolling black-outs, the NorthEastern blackout of 2004, and the South Florida blackout of 2008.

In a December, 2007 a report issued by Emerging Energy Research titled "Wind Power Strategies in the US 2007-2015" stated: "CSP production in the US and Spain expected to reach 7500MW by 2020 enough to power 6.75 million homes"

"CSP is the fastest growing utility-scale renewable energy alternative after wind power, with up to $20 billion expected to be invested in CSP over the next five years."

The long-term potential for solar technologies is even higher, as represented by the following chart (figure 2)**** from the Solar Wirtschaft (Germany).

Advantages of solar energy production are numerous. In addition to no carbon emissions, harnessing the energy from the sun pulls energy from a never-ending resource. The costs are fixed, and energy prices remain stable as there is no reliance on a fossil fuel. Solar is a complimentary technology to other forms of renewable energy, such as wind and biomass.

Midterm: What does it Take to Make CSP Cost Competitive

Many view solar technology to be cost-prohibitive, while this was true 20, even 10 years ago, thanks to innovations and improvements in efficiency from industry, the cost per kW/h is steadily decreasing. It is only through continued government support however, that the industry will continue to make investments in research and development, which will further reduce costs and bring them in line with electricity generation from traditional fossil-fuels.

A chart (figure 3) for CSP technologies again shows parity within the next decade with key productivity sources in economy of scale, increase of efficiency and the development of storage technologies. Funding is again assumed for eight years through the extension of the ITC.

Currently in the United States power from renewable energy sources accounts for less than 6.5% of the US energy consumption, of which solar is 1 %. However the US is showing one of the biggest growth rates with CAGR (compounded annual growth rate) of 36% from 2006—2011.

In a conservative market scenario, the overall US PV market will reach ~900 MW in 2012. Through an aggressive market scenario, the US market can more than triple, to almost 3GW of installed capacity by 2012. The aggressive scenario assumes a long-term (8 year) extension of the investment tax credit. This extension will allow for sustained manufacturing capacity expansion, as evidenced by companies like SCHOTT, who is investing $500 Million in a solar technology production facility in Albuquerque, NM. Strong demand growth must continue with minor supply excesses causing large price declines in line with unit subsidy rate declines. The 3GW market in the US compares to a 7.6GW world market.

Energy Cost Stability

With oil prices currently exceeding $140/barrel (6/30/08), and energy prices correlating with the price of oil, the need for fixed-price energy solutions is more important than ever. Solar represents fixed cost power generation. With more widespread deployment of CSP, through economies of scale and technological improvements, the costs of CSP power generation will continue to decrease. Currently, the cost to operate a CSP power plant is approximately 3 cents per kilowatt hour (not including the cost of amortizing the construction of the facility).

The following chart (figure 4)—labeled Exhibit 1-1, shows CSP deployment as it relates to the cost of natural gas.

Effective Legislation will Push CSP Development

Investment Tax Credit

There are key steps the Federal government can take to create a favorable climate for the deployment of CSP. First and foremost, a long-term renewal of the Federal Investment tax credit (ITC) is seen as an essential first step. Although the overall cost of the ITC extension is variable based on the amount of solar actually installed, independent analysis (GE Capital) has stated that the solar component (including PV) will most likely not exceed 2 billion USD over the 8 years. When compared with the job creation and the billions of dollars in investment by private industry, the payback on the 2 billion should not be difficult to recoup.

The following chart (Figure 5), labeled exhibit 1-16, shows how the ITC, along with other (global) legislation will spur development and deployment of CSP.

Federal Renewable Portfolio Standard (RPS)

Many states, including New Mexico, have enacted Renewable Portfolio Standards (RPS') which state that by a certain time, a certain percentage of electricity either generated—or consumed—in a State must come from renewable sources. Some even go further down by mandating a renewable mix where a certain percentage must come from solar. The Federal government could enact similar legislation, which would signal to the CSP industry a clear commitment, which would enable long-term investment. As a comparison, states with RPS' currently have 80% of the renewable energy projects in the pipeline compared to 20% of non RPS states (according to EER).

Feed in Tariffs (FIT)

A current stumbling block for the development of CSP is in negotiating power purchase agreements (PPA's) with utilities, who buy the renewable energy and then distribute it to customers. The FIT model, originally developed in the US, and successful deployed in both Germany and Spain (see case-study following) would create a Federal incentive to purchase energy from renewable sources. Since CSP is a utility-scale generator, this would ease the constraints of the utilities who are under pressure to deliver power to the end customer at competitive rates, but also are obligated, in many areas to purchase energy from renewable sources (from the RPS).

A national FIT is seen as one of the most effective means of rapidly growing the renewable energy market in the US.

Easing Land Management Restrictions

There is a current moratorium placed on new solar projects on Federal lands as the environmental impact of CSP power plants is currently being studied. While the industry understands and recognizes the importance of such studies, stopping all projects while commissioning an environmental impact studies is perhaps too far reaching. A compromise should be developed that strikes a balance between renewable energy and stewardship for the environment.

Transmission

Since CSP is currently not installed on a widespread basis, and the energy produced is therefore consumed in local regions—due to the extraordinary potential of the technology, a time will come when transmitting the energy to other regions will become necessary. In this regard, the Federal government can support utilities in creating a "smart grid" that will enable such transmission over the network of utility owned transmission lines.

National, Bi-Partisan Support for Solar

A recent (June, 2008) study conducted by the independent polling firm Kelton Research, demonstrated the tremendous support solar energy has across America. 94% of Americans, representing individuals across all political affiliations and geographic regions, support the development and use of solar power. Additionally, approximately 75% of Americans support the extension of the ITC and almost 80% feel that solar should be a "major priority" of the Federal government. When asked which one energy source they would develop if they were president, most respondents chose solar over any other type of energy generation.

Technological Breakthroughs on the Horizon

When speaking about electricity generation, you're speaking in costs. The cheaper the generation, the more widespread it will become. Critics of solar state that the energy produced is not cost-competitive with current methods, and it only works during the day.

Through support from the Federal government, private industry, will most likely overcome key technical hurdles in the technology, which will further reduce costs. Currently, the Heat Transfer Fluid (HTF) breaks down if it exceeds approximately 750 degrees Fahrenheit. If a suitable replacement can be developed the potential exists to heat the fluid to higher temperatures, improving the efficiency.

Additionally, in Spain, the first CSP plants that utilize molten salt storage units are currently being deployed. By storing the heat generated during the day, CSP plants could become a 24/7 operation without the need of a natural gas feed back-up.

Other advancements in the technology can be made through advanced coatings on the receivers, lighter and cheaper materials used in construction of the parabolic trough mirrors, and other areas of the power blocks. These advancements can be made if the CSP industry knows that a market will exist to deploy and utilize the technology that can be developed.

Case Study: A Model for Economic Development—Spain

After an early start as a world leader in solar energy, the United States lags behind several countries in both solar energy development and deployment. However, global warming and rising prices for fossil fuel are causing the United States to consider how it can regain world leadership in the generation of solar energy. Overseas best practices offer proven models for how the U.S. can increase solar energy production. One of the leading examples may be found in Spain, where the government has undertaken aggressive initiatives that have made that country one of the world's solar power leaders. Not only have these initiatives helped increase the amount of solar energy generated in the country, but they have spurred the development of Spain's solar power industry as well proving to be an economic stimulus and creating jobs.

There are some obvious reasons for Spain's leadership in solar power. For one thing, solar energy generation is simply the exploitation of one of the country's most abundant natural resources. As British and Scandinavian sun worshippers can attest, Spain enjoys more sunlight than any country in Europe. Yet, in many ways, this resource remained untapped until 2004, when the Spanish government issued Royal Decree 436, which made sweeping reforms to solar energy policy, creating a new system for renewable energy development and deployment, with its own regulatory framework.

The decree ended a regime of small steps toward promoting the use of solar power and instead initiated the adoption of bold policies that would strongly encourage the deployment of solar energy. These policies included grid connection and tariff reform, promotion of large-scale concentrated solar power (CSP) plants and later, solar panel mandates for new and renovated buildings. Their initial goal was ambitious— 30 percent of the nation's electricity to be supplied by renewable energy sources by 2010.

Grid Connection

Grid-connection is critical to the development of renewable energy anywhere. In Britain, for example, the Labour government refuses to remove obstacles to grid-connection, and solar energy development lags. Without some form of guaranteed grid access, it is difficult for companies other than the grid owners to develop large-scale solar power plants, severely limiting the number of companies who can enter the market.

In 2004, the Spanish government removed the economic barriers to grid-connection for renewable energy sources. With this single measure, large-scale solar power plants were guaranteed access to the electricity grid and a market was created for the solar energy generated at these plants.

Economic Incentives

Spain has made economic incentives, particularly feed-in tariffs, a key feature of its solar energy program. In 2002, Spain became the first European country to adopt a feed-in tariff of 12 euro cents for every kilowatt-hour supplied to the grid. In order to further accelerate the development of solar power the government passed a decree in 2004 that almost doubled the feed-in tariff for solar energy kilowatt hours, to 23 euro cents, and guaranteed these rates for 25 years. Instantly, large-scale photovoltaic and CSP generation were transformed into profitable business propositions as the 23 euro cents per KWh tariff was made specifically applicable to 100 KW to 50 MW plants. To keep the ball rolling, in 2007, the subsidies were raised yet again to 27 euro cents per KWh.

When combined with grid connectivity, these economic incentives made the development of solar energy in Spain practical. Planning and construction of solar generating plants in Spain accelerated, creating jobs and stimulating the economy.

Large Scale CSP

Though CSP is less well known than PV, since the 1980s CSP plants have reliably and cost-effectively generated large amounts of clean energy in California's Mojave Desert. Recognizing the tremendous potential that CSP offers geographic areas located in the world's sunbelt, Spanish policy essentially makes CSP fully equal to PV technology. With large areas that receive strong amounts of direct sunlight, Spain is very well suited for the development of CSP plants.

At the end of this year Spain plans to start operation of its first commercial CSP plants. The first plant, Andasol 1, will be the first commercial parabolic trough CSP plant in Europe. It will have a half-million square meter collector field and will be capable of supplying electricity to as many as 50,000 homes. This plant is the world's first to include thermal storage technologies that allow the plant to produce power at night. It does this by storing up to seven hours of energy in hot molten salt reservoirs. The heat in these reservoirs can be tapped to generate electricity after the sun goes down. Ultimately, this technology could enable solar plants to operate around the clock.

The Andasol plants are only the beginning. As of early 2008, five other Spanish CSP projects were underway, with a total expected capacity of 190 MW. Spain's tremendous CSP potential recently led my company, SCHOTT, to invest approximately $28 million in a new parabolic trough CSP receiver production facility in Spain.

Solar Mandates for New and Renovated Buildings

In addition to opening up the grid, providing aggressive tariffs to solar power generators and encouraging the development of both PV and large-scale CSP, Spain has undertaken another step towards a solar energy future. A new policy, introduced in 2006, mandates that all new and renovated buildings include either solar water heating systems or PV arrays. New homes must have solar heating systems capable of providing from 30 to 70 percent of their hot water, with the specific requirements to be determined by the building's location and expected water usage. These panels will not generate electricity, but they will help cut the demand for electrical power significantly. For non-residential buildings, such as hospitals and shopping malls, the standard is different. They are required to have PV panels that generate a portion of their electricity. The Environmental and Housing Ministries expect these mandates to bring energy savings of 30 to 40 percent for each building, and reduce carbon dioxide emissions by 40 to 55 percent.

In 2004 the Spanish government set a goal of 400 installed MW of PV and 500MW of CSP by 2010. Currently, it seems likely that Spain will easily exceed these goals before 2010. By 2007, about 600 total MW of solar generating capacity were installed, with more projects under construction and scheduled for completion in 2008 and 2009. In fact, four of the 13 largest PV power plants in the world are in Spain. Two plants in Jumilla and Beneixama each produce 20 MW and each deploys more than 100,000 PV panels. The two other plants are a 13.8 MW facility in Salamanca and a 12.7 MW operation in Lobosilla.

Lessons for the United States

The Spanish experience offers important lessons for the United States, and especially the American Southwest, given that its climate is similar to that of Spain. The first and most important lesson is that without bold long-term policies, solar energy generation will only grow in fits and starts. Unfortunately, U.S. federal solar energy policy legislation has been short-term, with incentives periodically allowed to lapse, providing developers with no certainty that these incentives will be renewed or changed. This deters investment, and does not persuade the public that Congress and the Administration are serious about renewable energy policy.

Solar power plants—like any power plants—are major commitments, expected to be operational for at least 30 years. These kinds of investments require long-term federal energy policies.

For example, the U.S. tax credit now applies to a range of renewable energy projects and affords a 1.9¢ per kilowatt-hour benefit for the first 10 years of operation for a renewable-energy facility. It also lapses at the end of 2008. So projects—solar, wind and other renewables—languish while their developers await Congressional action.

The U.S. could benefit from adopting other aspects of Spain's solar energy policy. If the U.S. instituted a national grid connection policy, developers would be better able to overcome the obstacles inherent to a federal system with multiple jurisdictions. Currently, these bureaucratic roadblocks slow down or completely stall the development of many large-scale solar energy projects. In addition, the U.S. could further spur solar energy development by mandating the installation of solar energy in residential or commercial buildings.

The United States, and especially its desert Southwest, possesses great potential for rapid solar expansion if policies akin to those of Spain are adopted. Many government officials, utility executives and citizens in the American Southwest already recognize this, and are taking action to develop the region's abundant solar resources, despite federal inaction. The Western Governors Association has set an ambitious goal of generating no less than 8,000 solar MW by 2015, and has recommended many regulatory and other public policy changes to promote solar and other renewable energy development. Early this year, Arizona Public Service announced plans to build the 280 MW Solana Generating Station near Phoenix.

Another lesson the U.S. can learn from Spain is that strong support for solar power provides many economic benefits. For instance, Spain's Ministry of Industry estimates that the solar and other renewable energy industries will create 200,000 new jobs by 2010.

The United States has found itself behind in the deployment of important technologies before, and found ways to catch up and secure world leadership. However, if our country adopts renewable energy policies similar to Spain, we can catch up just as we did with other technologies. And catching up will not just help the U.S. move beyond the use of fossil fuels and reduce its greenhouse gas emissions. Despite not having solar energy policies as aggressive as Spain's, the Solar Energy Industries Association (SEIA) estimates that 314 megawatts of new solar were installed in the U.S. in 2007, contributing $2 billion to the U.S. economy and creating 6,000 new jobs.

Solar Energy as an Economic Engine

Solar energy is domestic energy. The economic engine created by a powerful solar energy policy is multi-faceted. The most powerful component of the strengthening in the economy is in job creation. The University of California Berkley estimates "green jobs" will reach one million in the United States by 2020. These are high-wage manufacturing and professional jobs. In addition, there are a host of associated industries, such as plumbers and electricians that will also benefit.

It's forecasted that if the ITC is extended, 62,000 manufacturing and distribution jobs will be created—directly as a result of increased adoption of renewable energy in the first year of the extension.

This is job growth for Americans, by Americans, for an industry that will benefit America.

In addition to job creation, there are other economic benefits. Consumers will be able to combat volatile energy prices. Utilities will finally have a power infrastructure that can meet peak demand. Distributed solar can stabilize grids and offset expensive infrastructure upgrades. By 2020, the cost of generating solar power is forecast to become cost -competitive with fossil fuel energy production.

As an example, SCHOTT Solar, the company I represent, is in the construction phase of a large manufacturing facility in Albuquerque, NM. This plant will employ 1,500 people in the production of photovoltaics and receivers for CSP power plants. Over the long-term SCHOTT's investment in New Mexico will reach $500 million and the economic impact is forecast to exceed $1 Billion. But this growth will only happen if effective legislation is passed.

That's just what one company is doing in one community. There are other companies undertaking similar large projects, and many more that are ready to do so, once a clear commitment from the US government is established in the form of a long-term Investment tax credit. If the renewable energy credits expire, the impact next year would be more than 116,000 jobs either lost or not created according to SEIA and Navigant Consulting. Additionally, there will be more than $20 billion worth of investments that won't be made. And no doubt that that money, and those jobs, would go overseas. Considering the current economic climate of the country, these job losses, and investments moving overseas would be detrimental to the overall health of the nation's economy.

Solar as a Component of National Security

Currently the United States is reliant upon politically unstable regions of the world for much of its energy. According to the Energy Information Agency, two-thirds of the petroleum and 20% of the natural gas consumed in the United States is imported from other countries, and U.S. production of both is dropping while consumption continues to rise.

By installing solar powered power plants and the necessary infrastructure to transmit energy across the nation, states in the desert southwest could become an exporter of energy, helping economies in the region grow. Increasing energy consumption from renewable energy will stabilize energy costs and minimize wild fluctuations on the economy caused by volatile energy prices.

According to a study published in Scientific American (January, 2008) by 2050, solar power could end U.S. dependence on foreign oil and slash greenhouse gas emissions.

With sun shining all across the world, every country can develop solar energy as a means to create energy independence. Already, through solar, rural villages in South East Asia are benefiting from having electricity for the first time. Solar is scalable and deployable.

Summary and Recommendations

Renewable energy, specifically solar, represents tremendous potential for the United States. Through effective legislation, the United States can develop an industry with proven successes in Germany, Japan, and Spain. An industry that has the potential to create up to a million jobs domestically, reduce the country's dependence on foreign energy supplies, improve the environment for future generations.

- With the eight year extension of the Investment tax credit (ITC), an additional 62,000 jobs will be created. Up to a million will be clean-energy employed in the sector by 2020 according to UC Berkeley.
- By fostering developing of renewable energy, and specifically solar, costs will become competitive with fossil fuel based technologies by 2020.
- With the development of a National grid connection policy, solar project developers would be better able to overcome the obstacles inherent to a federal system with multiple jurisdictions. Currently, these bureaucratic roadblocks slow down or completely stall the development of many large-scale solar energy projects.
- With multiple GW of installed solar capacity, the US will be reducing its growing dependence on foreign energy source, which often come from politically unstable regions of the world.
- Strong support for solar will enable the industry to continue to make technological advances, including thermal storage, which extends the operating hours of solar power plants beyond daylight hours.
- CSP is a proven, reliable technology with a tremendous potential.

[Appendix documents have been retained in committee files.]

The Chairman. Thank you very much.

Mr. Wan, representing Pacific Gas & Electric, thank you for coming, and we look forward to your testimony.

STATEMENT OF FONG WAN, VICE PRESIDENT, ENERGY PROCUREMENT, PACIFIC GAS & ELECTRIC COMPANY, SAN FRANCISCO, CA

Mr. Wan. Senators, thank you for the opportunity to be here today and for this committee's leadership and commitment to advancing a clean energy future for the Nation.

My name is Fong Wan. As part of my role as Vice President of Energy Procurement at PG&E, I have the responsibility for overseeing the purchases of renewable energy.

PG&E has a long and accomplished track record on clean energy. We have contracted with about 40 renewable suppliers in the last 5 years. We are one of the Nation's largest buyers of renewable energy. Due to the large economy we serve, which is about 1 out of every 20 Americans, we have been fortunate to have the opportunity to meet with developers and technologies from all over the world.

We are committed to expanding our renewable supplies on an unprecedented basis. This includes a very sizable commitment to CSP. We have already contracted with four large solar thermal suppliers of various technologies that we can get into later. This amounts to about 1,700 megawatts. When these resources come online, they will represent enough power to meet almost 10 percent of our peak summer needs.

Moreover, we have stated a strong desire to pursuing even more opportunities in this area. That is because of this energy-producing availability at a time when our customers need it most and its relative cost effectiveness compared with many other renewable options. You have to take into consideration the time of the generation of CSP as well as theoretical potential. By National Renewable Energy Labs' estimate, CSP could, in theory, produce 7 times the energy needed to serve California.

I also want to provide our latest observation on photovoltaics. PV are making a strong and great progress for utility-scale applications. We also hope to be in a position to announce several contracts for utility-scale PV applications as well.

But given these advantages, it is reasonable to ask why the Nation is not seeing greater progress on renewables. As someone with daily experience in today's renewable energy marketplace, I would like to point out a few barriers. It is important to underscore that the Federal Government is uniquely positioned to help the country push past these obstacles.

The first is economics. Despite falling costs, CSP cannot yet compete on price with electricity produced by natural gas. We are confident that will change as the economies of scale are achieved and technology is refined. But in the interim, Federal production and investment tax credits are absolutely essential for continued progress.

Past experience shows a smart use of tax incentives makes tremendous difference in the pace and extended innovation and deployment of new renewable projects. The Federal Government should extend the PTC and ITC. It should also remove the ITC exclusion for regulating utilities. Utilities represent a significant source of potential new cost-effective capital investment. We should remove this handicap that effectively sidelines well-capitalized, motivated investors.

The lifetime of the extension is also critical. We believe the minimum commitment should be 8 years, as you have heard earlier. An 8-year extension would send a critical market signal. It would provide the assurance investors need to spur long-term R&D and allow the transmission to be built for large-scale CSP projects.

Without a longer term extension, we are deeply worried that developers will slam on the breaks and projects will be delayed, stopped, or prices will increase by as much as 30 percent to our customers. We, therefore, urge Congress to take action as soon as possible on the legislation it has before it.

Another significant challenge in bringing renewables online faster is transmission. Without new lines, we cannot get power from remote locations to the customer. Yet siting new transmission has become extraordinarily difficult. Mr. Chairman, you were exactly right when you recently said we need to be sure that first rules for planning, siting, pricing, interconnection, and openness of access are adequate.

Right now, it is not uncommon for a project to be stopped by a single stakeholder, and actions by some Federal agencies can also have major ramifications. We support BLM's desire for a comprehensive approach to solar projects in the Mojave Desert and throughout the West. We also appreciate BLM's commitment to continue to process those applications which already have been accepted. However, we hope that future CSP projects are not further complicated by the deferral on new applications.

A third challenge is integrating these intermittent renewable resources into our overall supply. One key is developing storage technology. We applaud Congress for including energy storage R&D program in legislation this year.

To summarize, in this time of high energy prices, a weak economy, and heightened focus on security, the Federal Government is uniquely positioned to provide clarity of vision and foster stable growth in this critical sector of the energy market. We encourage policymakers to address the challenges outlined, and we look forward to working with you to do so.

Thank you very much.

[The prepared statement of Mr. Wan follows:]

Prepared Statement of Fong Wan, Vice President, Energy Procurement, Pacific Gas & Electric Company, San Francisco, CA

Chairman Bingaman, Ranking Member Domenici, and Members of the Committee, I am very pleased to appear before you this morning on behalf of Pacific Gas and Electric Company to offer my views on the important role of concentrated solar power (CSP) as a clean, renewable source of energy. My name is Fong Wan and as part of my role as Vice President of Energy Procurement for PG&E, I have the responsibility for overseeing all of our renewable power procurement. At a time of historically high energy prices, increasing concerns over climate change and U.S. energy security, I commend the Committee for its leadership in addressing this important topic and for the continued commitment to and support for alternative energy that Chairman Bingaman, Ranking Member Domenici and others on this Committee have demonstrated over the years. The Energy Policy Act of 2005 and the Energy Independence and Security Act of 2007 have both helped to advance alternative energy, improve the overall energy efficiency of our economy and begin to dismantle barriers to ushering in a new energy paradigm for the 21st century.

Pacific Gas and Electric Company, headquartered in San Francisco, California, is one of the largest natural gas and electric power utility companies in the United States. The company provides natural gas and electric service to approximately 15 million people throughout a 70,000-square-mile service area in northern and central California. PG&E proudly delivers some of the nation's cleanest energy to our customers. On average, more than half of the electricity we deliver to customers comes from sources that emit no carbon dioxide, or CO_2, and an increasing amount comes from renewable sources of energy. In 2007, approximately 12 percent of our electric delivery mix was comprised of California-eligible renewable resources.[6]

PG&E is actively pursuing renewable generation resources on behalf of our customers for several reasons, including the following: first, it is what our customers consistently tell us they want; second, it furthers our efforts to meet the California renewable portfolio standard, which requires that 20 percent of our electric power be derived from renewable energy sources by 2010, a policy goal that PG&E strongly supports; and third, it allows us to better manage our future cost risk, on behalf of customers and shareholders, by taking volatile and rising fuel prices out of the cost equation for this portion of our generation.

PG&E has announced several contracts with wind, geothermal, biogas and solar developers. Solar thermal energy, the subject of today's hearing, is an especially attractive renewable power source because it is available when power is needed most in California—during the peak mid-day summer period. PG&E has entered into four solar thermal power procurement contracts totaling up to 1,737 megawatts of power, enough capacity to supply almost 10 percent of our peak summer needs. These include a contract with Ausra for a 177-megawatt facility in San Louis Obispo County, CA, a contract with Solel for a 553 megawatt facility in San Bernardino County, CA, and a contract with Brightsource Energy for 500 megawatts from facilities in San Bernardino County with an option for another 400 megawatts.[7]

We believe the potential for solar thermal technology, as well as other solar power technologies, is significant—and we are not alone. For example, a study prepared by the National Renewable Energy Laboratory (NREL) on the potential for concentrated solar power, or CSP, in California and the rest of the Southwest U.S. indicated that CSP in California could produce upwards of seven times the energy needed to serve the state. NREL also suggests that costs for CSP technologies could decline significantly, from approximately 16 cents per kilowatt-hour on average today, to approximately 8 cents per kilowatt-hour in 2015. The halving of the cost of this energy in seven years is premised on an assumption that at least 4,000 MW of CSP will be built by then—not just contracted for—to achieve "learning curve" benefits. For a comparison to another major energy technology development effort, cost estimates for advanced coal power generation with carbon capture and storage are on the order of 11 cents per kilowatt-hour.

We are also impressed by the progress being made in reducing the cost of photovoltaic technology and look forward to a healthy competition between CSP and utility-scale photovoltaics to meet the peak electric needs of California customers. We expect to announce a number of large, utility-scale photovoltaic projects in the near future. We think the competition between the two solar technologies will help our customers over time by bringing the cost of solar energy down.

As we move forward aggressively to deploy these renewable, clean, domestic energy resources, we recognize that challenges remain to fully realizing their potential in California, and across the nation. I will use the balance of my remarks to outline some of the challenges we see.

A. Extension of Incentives is Essential

As noted, while the cost of solar is anticipated to decline over time, competitive electric power pricing is perhaps the biggest current obstacle to more rapid and widespread deployment of solar and other renewables. Still a nascent industry in the U.S., solar has yet to reach economies-of-scale that will bring down the per-unit production cost to levels competitive with natural gas-fired plants.

One of the most important tools needed until the prices become competitive is the Investment Tax Credit (ITC). As a major buyer of renewable energy, Pacific Gas and Electric Co. is concerned that without proper tax incentives, there will be a significant slowdown in the development and construction of solar and other new renewable energy projects and technologies going forward, making it extremely difficult to meet the economies-of-scale required to drive down cost. An example of this phenomenon can be seen in the wind technology development experience. The expiration of the Production Tax Credit (PTC) in 2004, which is a key incentive for wind power projects, caused a 77 percent drop in installed wind capacity that year relative to one year prior, 2003. By comparison, in 2007, with the PTC in place, the wind industry enjoyed its best year ever when developers installed more than 5,000 megawatts of new generating capacity, more than twice the previous record. The ITC is expected to have a similar effect on the solar industry. The tremendous spurt of innovation, development and associated economic activity we have seen with solar, and in the renewable energy sector generally, could be squelched if these tax credits are not extended. Needless to say, the loss of this economic activity would occur just when the national economy, buffeted by the housing collapse and record energy prices, needs all the support it can get.

The federal government can therefore make a tremendous contribution by extending the ITC and PTC, as proposed in H.R. 6049, and removing the regulated utility exclusion associated with the ITC, as regulated utilities are in a position to provide financing for these capital-intensive projects. We believe the 8-year extension for the ITC included in the recent Senate package is the minimum amount of time necessary to reduce financial uncertainty, spur longer-term technology development and encourage fuller deployment of these projects. Given the long investment lead-times for CSP, an extension of 8 years would send a critical signal to investors to commit to these projects. In the absence of a long-term extension of the ITC, we are very concerned that the projects currently under development may be delayed, stopped, or priced in a way that will raise costs to electricity consumers by up to 30 percent. We therefore urge Congress to pass the provisions contained in H.R. 6049 as soon as possible, so that the solar and other renewable industries can have the certainty they need to make investments, create jobs, and make a positive contribution to meeting the nation's energy needs in an environmentally-sound and sustainable manner.

B. The Transmission Challenge

Another significant challenge we face in bringing renewable energy resources online faster is transmission. In California, for example, most large-scale concentrated solar power generating facilities are sited in remote desert locations, far away from the areas where the electricity is needed most. While siting in these areas may avoid some, but not all, issues associated with major power plant project siting, if we can't get the power to the customer, it's just a different kind of stranded asset. In fact, Senate Majority Leader Reid noted in his statement on June 17th before this Committee that the West alone will need approximately 7,500 miles of new transmission lines over the next decade to significantly expand renewable energy production.

Transmission siting is a multi-stakeholder process that is increasingly complicated in the case of interstate lines, due to multi-state regulatory requirements, myriad diverse stakeholder interests, and a lack of deference to a lead federal agency. It is not uncommon for a company to be far into the siting process and have a single stakeholder raise an objection which can stop the project's momentum 'in its tracks.' And actions by some federal resource agencies can also have major ramifications. We support the Bureau of Land Management's (BLM) desire for a comprehensive approach to solar projects in the Mojave Desert region and the west, and we appreciate BLM's commitment to continue to process those applications which have already been accepted. However, we hope that future CSP projects are not further complicated by the moratorium on new applications."

Mr. Chairman, I couldn't agree more with your statement at the June 17th Committee when you said, "[t]o get transmission built to carry renewable electricity, it's important to make sure that the transmission system in general is working well. We need to be sure that FERC's rules for planning, siting, pricing, interconnection and openness of access are adequate."

C. Integrating Intermittent Power Supplies

A third challenge we face is the ability to integrate an increasing amount of intermittent power resources into our generation portfolio. Solar is much easier to accommodate than wind in this regard, but it still poses challenges. A solution to this challenge lies in the ability ultimately to store excess power from these power projects, so that it is available when the project is not producing power—for example, on cloudy days or at night—and thereby smooth out the 'ups and downs' that are otherwise associated with intermittent power generation. Congress showed great foresight by including a thermal energy storage research and development program in Section 602 of The Energy Independence and Security Act of 2007, and hopefully this program will lead to improvements in the cost and effectiveness of such technologies.

The Path Forward

At PG&E, we are working cooperatively with policymakers, regulators, others in our industry and myriad stakeholders to help tackle these challenges. For example, California's utilities are working closely with state and federal agencies on the Renewable Energy Transmission Initiative to identify areas that will require transmission investments to bring on new solar and other renewable energy supplies. We are working in a broad coalition, consisting of more than 300 organizations to help support and advance the energy tax provisions contained in H.R. 6049. We are working to identify and support emerging renewable technologies, like CSP, biogas, and wave technology, to help bring down costs and integrate these technologies into the electric power and natural gas systems. And, we are investing more than a billion dollars in advanced meters to upgrade our electric grid to make it a dynamic, "smart" system that will allow us to optimize its performance and better integrate these new, clean technologies, including advanced transportation technologies like plug-in electric vehicles.

In these uncertain times of rising energy prices, a weakening economy, and increasing national security risks from dependence on foreign oil, the federal government is well positioned to bring certainty to the energy market through sound policies that send the right long term signals that will spur the innovation, development, and deployment of renewable, clean, domestic energy resources that are so desperately needed. The opportunity to expand the renewable power industry can only be accomplished by addressing the challenges facing it in an integrated, strategic fashion that blends incentives, standards, public sector investment, and other key mechanisms such as a price and market for carbon, a major step that will allow renewables to compete against the real costs of conventional sources of electric power generation.

On behalf of PG&E, I want to thank you for the opportunity to appear before the Committee today and I look forward to answering your questions. Thank you.

The Chairman. Thank you all very much for the good testimony.

Let me just start and ask a few questions, and then we will sort of take 5-minute rounds here and go back and forth on questions.

One issue that I am not real clear in my mind on is how you accomplish the storage requirement that we are all talking about here. I think, Mr. Morse, you said that the proposal or the project that you are doing with Arizona Public Service involves 6 hours of storage. I think that is what Mr. Nelson indicated they would like to see as well.

I don't know if you would like to explain how that storage is accomplished, or Mr. Andraka, if you would like to explain it? One of you who understands a little more of the technology involved, I would be anxious to hear it.

Mr. Morse. Let me take a stab at it. Imagine that you like coffee and you want coffee in your office at night, and you can't boil the water in your office. So what do you do? You boil more water in the daytime. You put it in a thermos, and you use it at night.

A CSP plant has an oil that runs through the plant that gets heated up to 700, 750 degrees Fahrenheit, and that hot oil makes the steam that runs the generator. If you expand the size of the solar field, then some of that collected energy goes not to make steam and run the plant in the daytime, but it goes into large, very large tanks of a fluid that has the heat capacity that can hold that high temperature.

Then at night, when the Sun goes down, you just run that fluid through a steam generator, make steam and run the power plant. So it is basically two large thermos bottles, one with hot oil that then runs and makes steam and then is collected in another tank, which is then heated up the next day.

The Chairman. Very good. This will allow you to extend the production of electricity for about 6 hours after the Sun goes down?

Mr. Morse. Correct. You could do it for 8 hours or 10 hours. You could run it around the clock. It is all the economics of the utility's needs. But it is feasible to make CSP a base load option if the economics made sense.

The Chairman. I think one of you indicated that by putting this additional storage capacity in, you do bring down the cost of the electricity per kilowatt hour. Is that the expectation?

Mr. Nelson. Yes, sir. We believe, based on the EPRI study, that the inclusion of storage has the opportunity to reduce the overall cost of the process. In addition to helping us address peak load needs in the evening, there is also some opportunities to allow that plant to startup earlier in the morning.

So, for instance, if you were hitting a winter peak early in the morning, that plant could be up and running and generating electricity even before the Sun generates enough heat to generate electricity. So it serves a couple of our needs in terms of peak load.

The Chairman. OK. Yes, go ahead.

Mr. Andraka. Senator, the labs are also working on advanced storage technologies. The technology that Dr. Morse talked about where you heat the oil and that, in turn, heats the storage fluid, we would like to extend that where you can directly heat the storage fluid in the field.

The issue there is the storage fluid right now freezes at about 200 degrees C. We have got new formulations we are working on that will stay liquid to a lower temperature so you don't have freeze issues in the field.

Mostly, we talked about trough systems today. An advanced technology is a tower system, where you are directly heating the salt in the receiver at a central point rather than throughout the field. So it gets rid of this distribution problem. We see storage as an inherent solution in towers as well as troughs.

One of the issues that also has to be addressed with storage is the contracts with the utilities that, in some cases, are regulated by the State that pay a certain amount at time of day. The storage increases the value of the electricity for the utility when they need it, but they need to be able to pass that value to the provider since we are looking at independent providers.

The Chairman. OK, let me ask one other question that is a little different from that.

Mike, you were talking there about—or maybe you weren't discussing it, but my understanding is that you are working on trying to develop an agreement with Kirtland Air Force Base on a power purchase agreement, something that would use power that might be produced at Mesa del Sol. Could you explain how that would work, how it would fit into the various other things we are talking about today?

Mr. Daly. Mr. Chairman, currently it is my belief that the base provides power to its tenants, which includes Sandia Labs, and that base negotiates a multi-year contract with WAPA, who acts as agent throughout, and both provide renewable power as well as traditional power on the wholesale market.

That contract might be—it is difficult to say what the renewable would be, but was 6 cents a kilowatt hour. It might be 7 or 8 cents going forward because of the increases. If DOE, as a tenant, decided that they wanted to take a portion of their load, 30 or 40 megawatts out of the total of 50 with the military being the balance, they are constricted for two reasons.

The first reason is they can't enter into a contract by law for more than 5 or possibly 10—there has been a recent event. It might be 10 years now power purchase agreement.

The second is that the proxy or the pricing under the OMB is that they have to match the lowest bid. Concentrated solar will never be the lowest bid today. But over that 30-year period if you have an extension of the power purchase agreement, or even 10, if you took into consideration an increase projected of fossil fuel cost, then they would be able to enter into that contract.

So there is a real OMB constraint and a contractual constraint preventing the biggest proponent of concentrated solar from actually being a buyer of concentrated solar for their own needs. This base and possibly other military installations have unique loads that really don't go down at night, too.

So we have actually worked a little bit on trying to straighten this out as a long-term possibility for DOE. But now that PNM has gotten the bulk of their requirement with the other utilities to 111 to 120 megawatts, we have a big enough plant. The original thought was that PNM's requirement wasn't big enough, and we would marry the DOE requirement and the PNM requirement to get the plant big enough because you have to do at least 120 megawatts.

I think long term it is both a policy and an opportunity. It is a great opportunity beyond having an additional generator backing up the entire installation here, which you could throw a switch and say they get the power first.

The Chairman. OK. Senator Domenici, go right ahead.

Senator Domenici. Mr. Chairman, I want to tell you I don't know how long you intend to continue, but I have no objection to your continuing if I am absent. I can't be here beyond 11:30. So that is 10 minutes, and you understand why. It has to do with my being present tomorrow.

Let me say that I have the cost of electricity by source per kilowatt hour, and I just want to make it a part of the record so it will be without a doubt.

We are getting there in terms of the solar we are talking about, the solar CSP. Obviously, we have got a ways to go, but I believe we will become more and more competitive if we get this—we can break it loose where it is not such an isolated job, but rather a contemplated and well-organized part of every utility that contemplates expanding so that they look at this first.

Not mandatory because we don't want to do that. But they look at it, and it is doable when we haven't put obstructions in. So I would like that to be in there.

Second, I did want to call to your attention, Mr. Chairman and this group of witnesses and our visiting Senator, the Albuquerque Journal seems to have gotten the understanding on what is going on. They have written about three editorials here in a row that impressed me with their understanding of the problem we have with crude oil dependency, which is slightly different than most people understand.

You know, we can't get out of oil for a long time because we can't get rid of our cars and our trucks. We are going to try, but that dependency is going to be there for a long time just because of the transportation and transportation users in place. You couldn't load up all the cars and put them out in the ocean on barges and ship them out, then we would leave an America that was dead on its toes.

It may be 30 or 40 years until dependence. That is terrible. I must say that every hearing I am going to be at, I am going to talk about a completely different view of our crude oil dependency. I think it is destruction we are talking about. It is destroying the economy—look—$500 billion a year to other countries for our crude oil is not an issue of, "Well, it is not too bad." The point of it is the whole economy is suffering, and we don't quite know why, and that is it.

The reason our economy is in terrible shape is because of this. We are sending too much of our lucre, too much of our would-be equity, too much of our money to others just for this one thing.

Now, what I am concerned about, Mr. Chairman, and you are going to be there a long time to try to organize, I think we have a lot of institutions in this country that aren't onboard as partners in trying to get where we ought to go.

For instance, there is a lead editorial here, "Killing Energy Options Will Leave Us In The Dark." It is about a $400 million investment that was going to take place of the kind you were talking about, Mr. Daly. It was going to be made in California, and it is about to die because the transmission line to make it operational is 23 miles long and must cross a State park.

There was a group who loves the State park so much that they have killed the project by saying you can't cross our park. We all are environmentalists. Maybe you on the record more than I. You more, but we all are.

But we have to have the environmentalists join up, too. They have to be concerned about the problems we are concerned about. This plant in this editorial must proceed. I am going to bring it up and have our committee find out why we stopped a $400 million solar plant with a 2-mile transmission line.

I don't see how that could hurt a State park unless there was a big cable right over a picnic area. You know, this is a big park. So, I just want to make that case and put that in the record, Mr. Chairman.

The Chairman. We will be glad to have all of that in the record.

Senator Domenici. Second, I would like for you all to know that the Bureau of Land Management is called a lot of things, and they are unable to do a lot of things we all want them to do. In this case, they are in the midst of a real argument, and we have to help them, I think.

They need some policy advice. They are left there with hundreds of applications, and they don't have a lot of help because they are not used to 100, 200 applications for a great big solar plant on BLM land. We don't want them to be the killer, right? We want them to be facilitators. But we have got to find out how they can help.

Did any of you run into this, just specifically, the BLM problem? Or is—yes?

Mr. Morse. I could make one comment that I think is relevant and that is causing a lot of the problems. There are three types of projects. There are real projects, a signed power purchase agreement with a commercial operation date—2012, 2011. Those plants have to be built, and if they want BLM land, their applications should be considered immediately.

Then there are projects that are maybe. Somebody applies for land for something that is still in their mind. They don't have a power purchase agreement. That is a real problem because you can fill up the queue with that.

Then there are the speculators. Let us just be honest about it. There are people with money who say that land is going to be worth something, I will put in an application. It is like grazing land without a cow. I think those are parts of the problem, and I think BLM overreacted.

Senator Domenici. OK. Let me just say I have been informed, and that is what this whispering was, they have changed their policy. So, maybe we had an impact and maybe we didn't. Your concern on the BLM is no longer there——

Senator Sanders. Do I understand that you are just telling us that they have lifted? They are not going to pose a moratorium. All right. We have some good news, good news.

Senator Domenici. That is good news, and I don't know how they are going to handle the vote. But anyway, I am grateful and that is nice that you called this meeting. That probably caused it.

[Laughter.]

Senator Domenici. My last point, Dr. Marker, I have never had a chance to thank you for your decision to come here. We are really glad to have you.

We hope that we have put in place the policies that will make your business thrive because that will be good for the country. In this case, we have got a winner. You win. The country wins. We win. So let us hope it happens.

But I would say to my friend, the chairman, there is probably not any better testimony than yours and yours, Mr. Daly—Marker and Daly—to support the proposition for the tax credit that we are looking for, that we want an 8-year extension. It is not the tax credit. What is it called? The investment tax credit.

There is probably not a better record than yours that clearly states that we don't have to worry about paying for this ITC. It is more than paying itself with growth that you are talking about.

Now I know my friend, Senator Bingaman, who I have great respect for his understanding and rationale, and he will answer that we are going in debt and how much more can we put on our grandchildren? I know all about that. I did that budget for 28 years for the Senate, and I did two of those balanced budgets, you know?

But, look, the point is we already have a policy of some sorts that will put this tax in place without an offset because it pays for itself. Look, you are telling us in your testimony that that is really true.

Anyway, I want to put this editorial in the record to remind us that transportation issues are important, and I want to close by saying to all of you we put into national law, you know, this business that I told you about in my opening remarks when you run into a wall, you pull it, take it away from the group that is arguing. You take it up with the secretary, and he sends it over to the commission, and they decide. We are catching all kinds of flak that we shouldn't be infringing upon these rights.

I want to tell you that that galls me. When I read in this Journal editorial that you are holding up a $400 million project in the State of California because an environmental group won't allow a 23-mile pipeline, I think the environmentalists have to join us. We need them.

But anyway, I think this hearing ought to be one that ends up saying right away this is very important, and we urge Americans to act like Americans on some of these issues and get rid of the parochialism of their cause and let it get in there and get solved, or the same thing is going to happen with nuclear.

I don't know if you are for it or not, but Senator Bingaman, at least he and I put in place what nuclear needs, just like we put in place what you need. Nuclear is going to come forward, and it is going to run into just what you are saying. We don't have enough trained people. We don't have enough engineers.

I never heard of solar engineers. But if that is a new doctorate or you have one? That is you?

[Laughter.]

Mr. Andraka. There is one.

Senator Domenici. One. Where did you get trained?

Mr. Andraka. Virginia Tech.

Senator Domenici. Virginia Tech. Do they give you a degree in solar?

Mr. Andraka. Mechanical.

Senator Domenici. Oh.

Mr. Andraka. I have become a solar engineer.

Senator Domenici. You can become that from mechanical. OK, well, that is what you are urging. You are going to get a seat in mechanical. How much does it cost to put one of those chairs in?

Mr. Daly. Too much.

[Laughter.]

Senator Domenici. I don't know. You are making a lot of money. You have got the best land deal anybody ever got.

[Laughter.]

Senator Domenici. I mean, I know your chairman. What is his name?

Mr. Daly. My chairman is Albert Ratner.

Senator Domenici. Yes, he is great. Whenever you worry about me not being on your side, you send him over.

Mr. Daly. I do. I do.

Senator Domenici. He is terrific. He is terrific. I understand him when he says he knows how to invest, and let him do that and we do some other things. I am all for it.

Mr. Daly. Thank you, Senator.

Senator Domenici. Anyway, you have got to be in this business—if you are going to run that big piece of land, you have got to be urging that people join together, right?

Mr. Daly. Correct, Senator.

Senator Domenici. Get some of their concerns, private concerns to join the cause.

Mr. Daly. Correct. I think the point there is no one person that can solve this. It is a collaborative effort with the utility, the States, and also the PRC has got some things they have to do to really allow us to do this.

Also the consumers have to recognize that green is going to cost them a little bit, too. It might be 25 cents a month on their electric bill, but we have got to get over that and go forward with it.

Senator Domenici. With that, Senator, Mr. Chairman, I am going to leave. I will let all these wonderful New Mexicans and you solve the rest of the problem.

The Chairman. Senator Sanders and I will solve the problem and report back to you.

[Laughter.]

Senator Domenici. You won't even need anything from me. Get it done.

The Chairman. All right. OK. Thanks for being here.

Bernie, why don't you go ahead with your questions?

Senator Sanders. OK. First of all, it is, in fact, very good news, I think, to hear from the BLM that they have withdrawn their proposal to establish a moratorium on new applications. Maybe it is coincidental that it may have something to do with this hearing, but we are delighted that that is the result.

But one point that I want to make, and I want to ask the panelists about that, is while we are delighted with the BLM's decision today, what I hear is that they have just a very few staff people who are trying to process these claims.

So all of you are talking about the need to reverse global warming, to create hundreds of thousands of new jobs, to make ourselves energy independent, and every one of you is saying that concentrated solar is going to be an important part of that process, and we have a bottleneck with two guys who are sitting there trying to process all of these applications.

So I am going to ask you a rather silly question, but do you think we should substantially increase the staffing at BLM so that they can process these applications in a far more vigorous way?

Mr. Morse, do you want to start on that one?

Mr. Morse. Of course.

Senator Sanders. Mr. Wan.

Mr. Wan. Absolutely.

Senator Sanders. You agree. Dr. Marker, do you agree with that? All right. OK.

One of the concerns that we have, and maybe one of you may want to go into it, I believe there are 130 applications in the pipeline. My understanding, and correct me if I am wrong, is not one of them has been approved yet. Is that——

Mr. Morse. That is what my understand is.

Senator Sanders. OK. So all of you are in agreement that we should substantially increase the staffing at the BLM to process these things.

Mr. Morse. I would also add I think that it is a complex issue. There are land management plans that have to be revised.

I will say that BLM is being very positive. They are trying to identify zones in the West that are ideal for solar energy. They are working with the environmentalists to make sure there are no wildlife corridors. They are going to do a programmatic environmental impact study. So they are trying to do the right thing, but they are understaffed. Certainly, more people would help, and they probably could use some other policy support along the way.

Senator Sanders. Mr. Chairman, the other question that I wanted to explore is the comparative cost of solar. By the way, yesterday I was in Nevada, and it is very hot in Nevada. You know that. am from Vermont. I didn't know that. But 110 degrees is very hot.

[Laughter.]

The Chairman. That is why we are in New Mexico.

Senator Sanders. You are right. I know that you have been talking, Mr. Daly, about working with your local base here. I was at Nellis Air Force Base, and I want to say that they have installed in a very rapid time, in about a 6-month period, not a concentrated solar plant, but it is the largest photovoltaic in the world.

It came in on budget. I think they did it in 6 months' time. It is producing more electricity than they anticipated. It is going quite well, and they have the capacity to produce even more. It was very nice to see a very positive relationship between the Air Force and the private sector and the environmental community. That is a very good omen, I think, for the military in general.

Let me quote—well, not quote, but my understanding, Mr. Andraka, is that according to a 2008 Sandia National Labs presentation, our costs for concentrated solar are projected to drop to 8 to 10 cents per kilowatt hour when capacity exceeds 3,000 megawatts. Is that your understanding?

Mr. Andraka. Yes, that is correct. That is based on a study done for the Western Governors Association, which also goes back to the Sergeant and Lundy report. DOE right now is funding an update to that Sergeant and Lundy report that will include dishes and update the cost estimates for troughs and towers.

Senator Sanders. Now tell me what I am missing here, but that sounds to me to be pretty reasonably priced electricity. The other point is that if you look at it over a 25-year period, it is not going to go up a whole lot, we don't think—unless Exxon buys the Sun or something.

[Laughter.]

Senator Sanders. That it is not going to go up too much. Is that right?

Mr. Andraka. The cost in a given plant will only go up with the labor——

Senator Sanders. Right.

Mr. Andraka [continuing]. Once the plant is put in place. Obviously, the plant depends on the cost of commodities, such as glass——

Senator Sanders. But my point is, and I think PG&E—and we talked about that as well, with Mr. Darby to help us, is that if you look at 8 to 10 cents a kilowatt hour, and you sign a 25-year purchase agreement, 25 years later, 10 or 11, whatever it may be, that is going to be pretty cheap electricity.

Am I missing something, or is that correct, Mr. Andraka?

Mr. Andraka. That is correct. Usually these contracts will have some inflation escalator built in, but we see the cost of natural gas going up a lot more rapidly.

Now in the last decade or so, the cost of natural gas was seen as cheap as far as we could see it, and that has been one of the impediments to rolling out——

Senator Sanders. Right. But compared to the volatility of gas or oil, the Sun is going to be reasonably stable?

Mr. Andraka. Yes. Yes.

Senator Sanders. Mr. Wan.

Mr. Wan. I think I would like to offer that even with conventional power plants, as they are facing the increased costs, there are still cost of raw materials, of the concrete. The conventional power plants are facing an increase of roughly 20 percent in cost. I would expect renewable generation, including CSP, to be facing that type of cost pressure.

Senator Sanders. For construction?

Mr. Wan. For construction, absolutely. That is without the impact of natural gas prices. Once it is in, it is minimal replacements on the mirrors as well as the tubes, and you are absolutely right.

Senator Sanders. Maybe Mr. Morse or anybody else would want to comment? I mean, there are a lot of things and you have said it all about why we should be moving forward in an aggressive manner for concentrated solar. But cost may, in years to come, be one of the reasons. It may be a very competitive product.

Mr. Morse.

Mr. Morse. In fact, Arizona Public Service, when asked why are they spending a little more for CSP today, the answer was it is a fixed price. It will never rise. They have no idea what natural gas will cost next year, 5 years, 10 years, if we will be able to burn it. It is a hedge against that.

Not only is the price fixed, when the debt is paid, there is no fuel cost—unless Exxon buys the Sun. If there is no fuel cost, it is a few cents a kilowatt hour. It becomes a clean cash cow. It is like a hydro plant that is paid for, but it is clean. Any utility that owns one of these, like PG&E, will be very happy that they have it.

So it is a very sound investment. The fact that it is a little costly now, that is what the ITC is helping to deal with. But as an investment, it is the most prudent investment that the country can make, and a lot of renewables have this same factor. They don't have a fuel price issue.

Mr. Wan. I would like to just offer a few numbers to illustrate the point. For example, if a natural gas combined-cycle power plant today is producing energy at roughly 10 cents, probably 40 percent of that is from the fixed cost of construction or maintenance, and the other 60 percent or so will be coming from the price of natural gas.

So Mr. Morse's point earlier, that 6 cents could be quite volatile in the future, and in his example, if CSP cost more than 10 cents today, but you are essentially locking in that number. Maybe it would be 11 or 12 cents. But you are far safer, from a portfolio management perspective, to have some CSP in your portfolio.

Senator Sanders. Mr. Chairman, may I ask——

The Chairman. Sure.

Senator Sanders. OK. We have been focusing, appropriately enough, on the huge potential of concentrated solar. When I was at Nellis, I was also impressed by the use of photovoltaic. So we talked to some people in the Energy Department a couple of months ago who surprised me by saying that they expect photovoltaic costs to also go down very substantially in the coming years and for people to install them on their rooftops.

California has had good success, New Jersey. Germany certainly has gone off the wall on this. What do you guys see as the potential of photovoltaics in the energy mix in this country?

Mr. Wan. As I mentioned earlier, we are in the final stages of negotiations with several of the PV developers, including the one at Nellis. The numbers they have all asserted to us are surprisingly low. We have thought for quite a while that CSP would be the future without any question, and I think PV is going to give CSP a run for the money. I think that is ultimately in the best interest of our consumers, for all Americans, as you have more competing industries and technologies to bring down the cost.

I think the key to photovoltaic today is not necessarily the efficiency of the panels or cells or to bring the cost down. It is actually the worldwide logistics of parts as well as the installation. That is why you see large-scale installation such as at Nellis.

Senator Sanders. The beauty of PVs is that while in the Northeast we are not going to be installing in the near future concentrated solar, we can be heavily utilizing PVs.

I mean, Germany is not an optimal location for solar exposure. It is nowhere near what it is here, for example. But New Jersey and other States, and we are trying to do this in Vermont, is to expand the use of PVs.

So you see great potential there as well?

Mr. Wan. We see great potential. We are very hopeful.

Senator Sanders. Yes, Mr. Nelson.

Mr. Nelson. Senator, I would like to give you our perspective on that. We actually are going to have another renewable energy RFP go out in the August timeframe. With that, it is going to be more of a general renewables RFP that we expect will include proposals related to PV, to biomass, to geothermal. We believe that PV plays a significant role in the renewable energy mix.

At this point in time, we believe that is more of a distributed role. Not just distributed in terms of rooftops, but megawatt scale located throughout the system to support distribution and that sort of thing. We are not yet, at this point in time, convinced that hundreds of megawatts in a single location is the most appropriate use for PV due to the large ramp rate associated with that when cloud cover comes over.

Senator Sanders. Mr. Andraka.

Mr. Andraka. I would like to echo the comments of Mr. Wan on the photovoltaics. One of the barriers is the interconnects, the field installation. One of the areas the laboratories are working on is modular systems, where you have got the PV built into a module and combined with an inverter. So it is an AC module that bolts into place to simplify and reduce the cost of the field installations.

Senator Sanders. Mr. Chairman, thank you very much.

The Chairman. Thank you very much.

Let me ask one issue that is always at the forefront here in the Southwest and New Mexico is water. Is there a significant issue of water usage in connection with a CSP facility like the one you are doing in Arizona or like the one PNM is proposing to do here?

Mr. Morse. It is all a problem of the French. Mr. Cugnot burdened us with the need if you have a heat engine, a steam turbine, you have to cool. You have to condense the fluid. A CSP plant, a trough plant or a power tower, is the same as any conventional plant. Coal, nuclear, doesn't matter. So we use the same amount of cooling water per megawatt.

We could go to completely dry cooling, air cooling. The cost would go up about 10 percent, and the performance, worst of all, would go down. During the hottest peak time, we would lose more performance. The solution that a lot of CSP developers are looking into is hybrid cooling, where we do dry cooling except during the very peak time, the hottest time, when we use a little bit of cooling water.

So I think that if the Southwest wants to have its future plants without water cooling, it can be done, and it is a matter of both a little cost and a little performance.

The Chairman. Yes? Mr. Nelson, do you have a point of view on that?

Mr. Nelson. Yes, sir. I would point out that our RFP actually requires that the vendors propose both dry and hybrid cooling options. We allow them to propose a wet cooling option as well, but we believe that the dry or hybrid is probably the way we need to go because of sustainability reasons. Water is a significant resource for us.

The Chairman. On this issue about the cost per kilowatt hour, as I understand, one of these concentrated solar operations—concentrating solar power plants, one of the great advantages is that it is producing the power when you are at your peak as far as demand. At least in the summer when people are using their air conditioners, you have got the most power going into the grid at that time.

How does that factor in? First, tell me if that is true as one of the advantages that concentrated solar has over wind power, for example. You just can't predict when the wind will blow.

Second, how does that relate to this cost per kilowatt hour issue? Because you have real-time pricing being factored in by Public Utility Commission, where you are charging a higher price per kilowatt hour during the peak periods, and you are producing this power even if it is more expensive during peak periods. How does all that fit together?

Mr. Nelson. We currently in New Mexico do not have peak or time of day pricing here, although that is certainly an option for the future. Solar power, however, does more closely meet our peak than wind power does, although it is not a perfect match. That is why we believe storage has the option to get us through our true peak, which is generally when people are coming home at the end of the day, turning on their air conditioning, turning on the TVs, the computers, the stoves, and that sort of thing.

So storage can help us get from that 5 o'clock to 8 o'clock at night type peak. But solar does match it much better than wind. What we know here in New Mexico is we have a great wind resource. However, the windy days are rarely the very, very hot days here in New Mexico.

We generally see our greatest wind resource in what we call the shoulder months, the springs, the falls. At nighttime, where our load demand is significantly less. So, again, from that perspective, solar better matches our peak.

The Chairman. Yes.

Mr. Andraka. We do see quite a diversity in the load profiles. For example, the Phoenix area, the peak extends maybe to 8 o'clock or 9 o'clock at night, and their request for proposals specifically require storage, and the plant that Dr. Morse proposes includes 6 hours of storage because of that requirement.

At the same time, the Southern California plants are not requiring storage. Their peak is a closer match to solar, still not a perfect match. But they also have significant wind resources, and those wind resources really start picking up every evening with the breezes off the ocean. So, in California, we see a portfolio of renewables in meeting that need.

Now as the solar resources are exploited and become a much greater portion of the grid, I think we will see more and more in California the need for storage, at least for the intermittency as used, if not the shifting.

The Chairman. OK. Yes, Mr. Wan.

Mr. Wan. Getting back to your question on the pricing, what you were hearing earlier is the average price. In California, we do pay a time of the day, time of the use value pricing. So that means if you are producing at the peak of the day, you can get up to 150 percent of that price that you heard earlier.

Similarly, if you are producing like wind in the middle of the night, you may get a lot less than that pricing. CSPs, we find, are critically dependent on the higher pricing for the power they produce over the peak of the day.

The Chairman. So you think that the time of day pricing that you have in California makes a lot of sense as you go to something like more power production from solar?

Mr. Wan. Absolutely.

The Chairman. Right.

Mr. Wan. Absolutely. Appropriately also pays the other technology less when they are not producing at the right time.

The Chairman. Mike, did you have a comment?

Mr. Daly. I just think, as I am listening—I can't speak for the other colleagues here—the not having a strong research program in solar storage, while it has been implemented in two or three situations in a solar tower, is the Achilles heel of the solar—concentrated solar business because if we can't store it, then we have to build gas plants to back it up anyway, and you are double-building capacity, which is going to hurt the rate bearer.

So as you look at research dollars, the solar storage really makes a big difference in this business. Then the second step is getting the local PRC to start doing time of day pricing so you start having market-driven consumption, which will add to the efficiency. But that is more of a local issue.

But storage is really important, and while it has been implemented, it is not widespread proven technology. It is a credit risk, and it is expensive to do it.

The Chairman. Senator Sanders, do you have other questions? Go ahead.

Senator Sanders. Just maybe one more, Mr. Chairman. To reiterate, I think this has been a very productive hearing, and I want to thank you for calling it and all of the panelists for being here.

My last question has to do with your views on a national renewable portfolio standard. My understanding is that New Mexico is moving to 20 percent by the year 2020, Nevada 15 percent by the year 2013. Other States have gone forward more or less significantly. We tried, and we lost by one vote, was it?

The Chairman. It was close, yes.

Senator Sanders. One or two votes. It was 15 percent, I think, right?

The Chairman. It was 15 percent, right. By 2020.

Senator Sanders. Right, 15 percent by 2020. What impact would a national renewable portfolio standard have on the solar industry?

Mr. Andraka.

Mr. Andraka. As we mentioned earlier, the primary resource is in the Southwest United States. So I think our States would be at a distinct advantage in taking advantage of the solar resource. Your meeting a few weeks ago on transmission capability needs to also look at transmission across the country, totally different grid technology to bring this resource to other parts of the country.

Senator Sanders. We have got a lot of work to do on that.

Mr. Andraka. Yes.

Senator Sanders. Yes, Mr. Wan.

Mr. Wan. PG&E has been a long supporter of the adoption of a national renewable strategy. We believe it should be uniformly applied. I think the most important part is that a national RPS policy will actually accelerate development and bring down the cost with greater scale and just move us so much faster and so much closer to the clean energy world that we would like to see.

Mr. Morse. I would like to add a word of caution, however. I think the word “portfolio” has to be looked at carefully. It could be a wind standard very easily because it could go to the lowest cost, and at the moment, that is the lowest cost. But the value for the utilities has to be considered.

So I think, No. 1, it should not harm what may be more aggressive State RPSs or RPIs, whatever the plural is. I know that there is hesitancy about carve-outs, but a lot of solo ones exist because they had a percent had to be solar. I know that that is a contentious issue. But I think if you do have a national portfolio standard, please be very sensitive to how it will be implemented.

For the life of me, I never understood States who pass a State RPS for generation without thinking a bit about transmission or enough about transmission. So think about if you were to require X percent new generation from renewables, how on earth is that going to get to the load centers?

Senator Sanders. Thank you, Mr. Chairman.

The Chairman. All right. Thank you all very much. Thanks for your testimony. I think it has been very useful. We will take all this information and try to put it to good use.

Thank you. That will conclude our hearing.

[Whereupon, at 11:59 a.m., the hearing was adjourned.]

APPENDIX: RESPONSES TO ADDITIONAL QUESTIONS

Responses of Michael Daly to Questions from Senator Domenici

Question 1. I understand the DOD will be contracting for this power. Now that they can enter into 10 year contracts for renewable energy instead of just 5 years, will the Defense Department extend its contract with Mesa del Sol?

Answer. We met with DOD in New Mexico for the past 2 days. They are hesitant to pay a premium for concentrated solar, i.e. greater than 9 cents per kilowatt hour. They have agreed to work with us and the City of Albuquerque to look into developing a plant possibly in conjunction with the PNM RFP. It is our position to combine the load of the base and the New Mexico utility companies which will result in the lowest possible costs for solar thermal. While a longer contract is preferable, 10 years would be sufficient.

Question 2. How are you handling the project's thermal storage?

Is there room to expand at your site between Sandia and Kirkland AFB?

Answer. We are proposing 3 hours of solar thermal which is consistent with the PNM RFP. The entire site between Mesa del Sol and Kirkland AFB is approximately 2,700 acres which is more than sufficient to accommodate an excess of 400 MW of concentrated solar. We are proposing to start at the northerly end of the site and development to the south. To the extent the full site was developed, we would have to coordinate with the Base to make sure we wouldn't impact the rocket sled or other Sandia missions.

Transmission

Question 3. In the 2005 Energy Policy Act, Congress sought to address the critical issue of transmission siting through the National Interest Electric Transmission Corridor process. Even though these provisions haven't been fully implemented, and no line has been sited pursuant to EPAct, the NIETC process has proven controversial. Still, everyone here today has highlighted the critical need to bring more transmission on line to transport these renewable resources to load. Just this past weekend, the Albuquerque Journal ran an Op-Ed criticizing environmental groups—who want the "green" power but not the infrastructure that goes with it—for opposing needed transmission lines.

What more should Congress do in this important area? Some have called for Congress to provide FERC with exclusive jurisdiction to site new transmission for a renewable project. Please comment.

Answer. Where transmission corridors are required to promote renewable energy and there are not other economically viable paths, FERC should have power condemnation after conducting public hearings.

ITC

Question 4. One of the most important issues facing the solar industry today is that the tax credits we passed in a bipartisan manner are set to expire. We must enact a long-term ITC extension as soon as possible. However, for the first time in the renewable tax credit history, the House Majority is insisting that the tax credits be "offset" by tax increases on other industries.

Many of you have submitted testimony highlighting the tremendous economic boost the ITC provides the solar industry. According to Dr. Marker from SHOTT, solar capacity additions in 2007 contributed $2 billion to the U.S. economy, creating 6,000 new jobs. And, it's forecasted that if the ITC is extended, 62,000 manufacturing and distribution jobs will be created.

Do you agree that these tax credits "pay for themselves" and therefore don't need to be paid for by raising taxes on other industries? Is the renewable industry concerned that this new "pay for" requirement can set a troubling precedent in that offsets will be required each time the existing tax credits expire?

Answer. The ITC should be extended to 8 years at a minimum, offsets should not be required to pay for the ITC as an impact on the economy will more than pay for any tax credits extended.

Solar Resource Potential

Question 5. We currently have just over 400 MW of CSP installed capacity in this country, at a rate of about 16 cents per kilowatt-hour. A recent study suggests that solar energy could grow to 10% of the nation's power by 2025. Do you agree with that assessment? If so, what percentage will CSP contribute as opposed to Photovoltaic? Also, how long will it take to get solar power costs on parity with conventional power sources?

Answer. I agree with the assessment. We believe that CSP's will represent the majority of the grid scale solar generation and that photovoltaic is better suited to distribute energy primarily due to the fact that there does not seem to be a short term solution to storage for photovoltaic's vs. concentrated solar which has commercially viable options that can carry a plant's generation capacity through peak demands in many markets.

Responses of Frederick H. Morse to Questions from Senator Domenici

Question 1. I understand that you have a new 280 mw trough CSP project under development in Arizona and that this project will have a thermal storage component. How much cost does that add to the plant? Does the addition of storage capacity improve the project's economics since you'll be able to increase power generation over a longer period of time? Why didn't the recent Nevada CSP plant include thermal storage?

Answer. The addition of six full load hours of thermal energy storage at the 280 MW Solana plant will increase the capital cost by approximately 15-20%. However, by adding thermal energy storage, the cost of the electricity generated by this plant will decrease and the value of that electrify to Arizona Public Service increases significantly because it can be used when it is most needed. I do not know why Nevada Solar One did not use thermal energy storage but perhaps the utility wanted a peaker and did not value enough the ability to shift the electricity generation to other times.

Question 2. The site location for the Abengoa plant is located near an existing transmission line. Would you be able to proceed with this project absent that line?

Answer. No, not at this location. If we were to site Solana at another location in order to access transmission, we would need to determine if the added cost of wheeling or additional interconnection would make the project uneconomic.

Question 3. Do you expect your plant to take the typical 7-8 years to be completed? How long did the recent NV plant take to come on-line?

Answer. Solana is now expected to come on line in 2012, provided that the ITC is extended very soon. I do not know the time it took for Nevada Solar One to come on line.

Transmission

Question 4. In the 2005 Energy Policy Act, Congress sought to address the critical issue of transmission siting through the National Interest Electric Transmission Corridor process. Even though these provisions haven't been fully implemented, and no line has been sited pursuant to EPAct, the NIETC process has proven controversial. Still, everyone here today has highlighted the critical need to bring more transmission on line to transport these renewable resources to load. Just this past weekend, the Albuquerque Journal ran an Op-Ed criticizing environmental groups—who want the "green" power but not the infrastructure that goes with it—for opposing needed transmission lines.

What more should Congress do in this important area? Some have called for Congress to provide FERC with exclusive jurisdiction to site new transmission for a renewable project. Please comment.

Answer. More transmission is absolutely essential for moving renewable energy generated electricity to load centers. Its absence will certainly limit how much renewable energy can be developed. There is a clear need for some way to deal with the conflict over who can decide what lines get built and where. However, concerns about reliability and economics militate against granting renewable suppliers such a mandate. Additionally, the Energy Policy Act of 1992 prohibits discriminatory use of the transmission system and therefore requires all generation resources equal access and use of the transmission grid.

The transmission system needs to be designed to integrate all sources of generation, in addition to renewable resources, into the entire system and managed as a whole, to be efficient and maintain reliability. To achieve a high level of penetration of renewable resources in an area, such as CSP, it must be able to interact with other areas to maintain the required real-time load-generation balance. That interaction requires sufficient transmission capacity between the areas, and this consideration alone will require expanding the system.

If giving FERC exclusive jurisdiction to site new transmission for renewable envy projects can withstand the likely challenge by the states, then this should be very helpful. For CSP, perhaps FERC's jurisdiction could be limited to siting new transmission from the solar zones that are under study by the BLM, the Western Governors' Association and the California Renewable Energy Transmission Initiative and to projects above a minimum size, say 250 MW.

ITC

Question 5. One of the most important issues facing the solar industry today is that the tax credits we passed in a bipartisan manner are set to expire. We must enact a long-term ITC extension as soon as possible. However, for the first time in the renewable tax credit history, the House Majority is insisting that the tax credits be "offset" by tax increases on other industries.

Many of you have submitted testimony highlighting the tremendous economic boost the ITC provides the solar industry. According to Dr. Marker from SHOTT, solar capacity additions in 2007 contributed $2 billion to the U.S. economy, creating 6,000 new jobs. And, it's forecasted that if the ITC is extended, 62,000 manufacturing and distribution jobs will be created.

Do you agree that these tax credits "pay for themselves" and therefore don't need to be paid for by raising taxes on other industries? Is the renewable industry concerned that this new "pay for" requirement can set a troubling precedent in that offsets will be required each time the existing tax credits expire?

Answer. I am well aware of the opposing views on if and how the ITC extension should be paid for however I am not in a position to comment on that. Solar energy has become another engine of the US economy and the extension of the ITC will allow solar to continue creating more jobs, contributing to economic growth and helping to reduce carbon emissions. Failure to find a way to extend the ITC now will result in the loss of over 25,000 jobs and the loss of the economic benefits of about $18 billion in investment in the 4, 800 MW of CSP plants now under contract but not financeable without the eight year extension of the ITC. And this at a time when America is trying to stop the loss of jobs and address serious energy concerns. I hope that a way can be found to pass the renewable energy tax credit extensions. To stop solar in its tracks seems like the wrong outcome for America.

Solar Resource Potential

Question 6. We currently have just over 400 MW of CSP installed capacity in this country, at a rate of about 16 cents per kilowatt-hour. A recent study suggests that solar energy could grow to 10% of the nation's power by 2025. Do you agree with that assessment? If so, what percentage will CSP contribute as opposed to Photovoltaic? Also, how long will it take to get solar power costs on parity with conventional power sources?

Answer. A study done by NREL estimates that CSP could provide nearly 120, 000 MW of capacity by 2050. Although this represents only about 6.8% of the nation's power, it would be a significant percentage of the power needed by the Southwest. The NREL analysis assumed all CSP power stayed within one state from where it was generated. This assumption exemplifies the fact that the growth of CSP power is limited by transmission. If DC transmission were available to move power from the SW to the East, CSP's potential would be greatly expanded. From a national perspective, solar power could be particularly beneficial to the eastern half of the country because the sun shines in the West throughout the East's evening high demand period. Because PV can produce electricity throughout the country, its potential could be higher than that of CSP. Regarding the time it will take for solar power to achieve parity with conventional sources, I believe that if the ITC is extended for eight years, CSP will be cost competitive with convention power by the end of those eight years. However, both CSP and conventional power will increase due to increased commodity prices.

Responses of Charles E. Andraka to Questions from Senator Domenici

Question 1. According to Sandia, it is possible to lower the CSP costs from today's 16 cents per kwH, to 12 cents in the near-term and even 6 cents in the long-term. How long do you estimate it will be before CSP drops to 12 cents? 6 cents?

Answer. The cost reductions for CSP technology, as for any emerging power system, are largely dependent on deployment. The costs to which you refer are from a 2003 report done by the DOE for the National Research Council,[8] commonly referred to as the "Sergeant and Lundy Report." While the costs of all technologies, especially fossil-fuelbased, have increased in the interim, this study reports that the cost reductions for trough and tower systems will result about equally from three sources: deployment, learning-curve cost reductions related to financing; and R&D of new components. The study indicates that cost reductions as low as 10 cents could be achieved with as little as 3 to 5 GW of deployment of CSP systems. (Note: there are currently 4 GW of CSP projects planned for deployment in the southwest.) The lower end of the scale, in the 6 cent range, requires aggressive technical advancement as well as deployment, as stated in the Sergeant and Lundy report:

> The specific values will depend on total capacity of various technologies deployed and the extent of R&D program success. In the technically aggressive cases for troughs / towers, the S&L analysis found that cost reductions were due to volume production (26%/28%), plant scale-up (20%/48%), and technological advance (54%/24%).

The net of these aggressive reductions reaches the 6 cent range. The original Sergeant & Lundy study is currently being updated and new number should be available late in the Fall. It should also be noted that the 6 cent range was seen in 2003 as the level needed to be competitive with conventional fossil technologies. The current market conditions will likely raise this competition level closer to 10 cents.

While it is not possible to say exactly when the cost of CSP systems will drop below 12 cents per kWhr, it was reported that a recent large US trough project was bid at 14 cents per kWhr.

Question 2. To date, there are no Power Tower CSP projects in this country, although an 11 MW project just came on-line in Spain. What are the advantages/disadvantages to this technology? Why aren't we seeing any Power Tower projects developed in the U.S.?

Answer. An 11 MW power tower project was built in Spain because of the favorable CSP incentive structure that will pay 47 UScents/kWhr for power from CSP; because a large company (Abengoa Solar) recognized a worldwide business development opportunity; and because of an additional government subsidy for the project.

As noted in the Sargent & Lundy report, the power tower has a longer-term cost advantage relative to solar trough systems. This is largely due to the fact that a power tower is a higher efficiency system and can more readily integrate thermal storage, thereby improving the capacity factor (yearly hours of operation) and potential value to the utilities.

So far we have not seen the deployment of power towers in the U.S. because of the perceived technical and financial risks. Due to the large initial capital investment needed for these projects, it has been difficult to obtain power purchase agreements and financing for higher risk projects. However, there is one power purchase agreement for a power tower in California between Pacific Gas & Electric and BrightSource energy. At least two other companies, eSolar and Solar Reserve, are actively developing projects in the southwest U.S.

The fact remains that we must find a way to enable the deployment of higher-risk, technologies at the utility-scale of solar power.

Question 3. I understand that Sandia is studying the new 64 MW CSP project in Nevada. What are you learning from that project? Why didn't that project include a thermal storage component?

Answer. The Nevada Solar 1 project was funded by private industry. Sandia is familiar with the technology and has worked with the developer, Solargenix, in the past but we do not have access to the data on the operation of the plant.

This system does not include thermal storage for two reasons: the utility, Nevada Power, did not value thermal storage to increase the capacity factor of the plant and they wanted the lowest cost solar option.

It should be noted that, even though two trough plants with storage are under construction in Spain, thermal energy storage for trough systems is not well established and some utilities and financial institutions consider it to be a higher risk technology. The announced 280 MW trough plant for the Phoenix area also includes thermal storage. While extending troughs to include storage is a current area of intense DOE and industry research, the potential for storage in towers appears to be much greater. This is because of the higher temperatures inherent in tower systems, which leads to less storage volume for a given level of storage. In addition, current storage proposals for troughs have a separate storage medium, requiring an expensive heat exchanger, whereas towers store the energy in the operating fluid (salt). We expect that current DOE development will support an increase in the number of plants, both towers and troughs, that incorporate thermal storage.

Question 4. Currently, a CSP project requires approximately 5 acres of land per megawatt. Is there room for improvement?

Answer. There is a small difference in the area required for a trough, a tower, and dish/Stirling systems. The size of the solar field for all three of the technologies is driven by shading issues in the morning and evening, as well as mid-day in the winter when the sun is lower in the sky. Any reduction in the area required for the concentrators will be incremental because it depends on the fundamental optics of concentrating sunlight. The plant concentrator area could be reduced by tradeoffs that the developer would make between energy production at certain times of year and plant foot print.

Transmission

Question 5. In the 2005 Energy Policy Act, Congress sought to address the critical issue of transmission siting through the National Interest Electric Transmission Corridor process. Even though these provisions haven't been fully implemented, and no line has been sited pursuant to EPAct, the NIETC process has proven controversial. Still, everyone here today has highlighted the critical need to bring more transmission on line to transport these renewable resources to load. Just this past weekend, the Albuquerque Journal ran an Op-Ed criticizing environmental groups—who want the "green" power but not the infrastructure that goes with it—for opposing needed transmission lines.

What more should Congress do in this important area? Some have called for Congress to provide FERC with exclusive jurisdiction to site new transmission for a renewable project. Please comment.

Answer. There are two approaches to address the nations growing electricity demand, through centralized generation and through distributed generation. Centralized generation includes coal, nuclear, wind farms, concentrated solar power plants, photovoltaic arrays, geothermal energy and other generation sources that produce electricity on the megawatt through gigawatt scale. Most of this centralized generation connects to the grid at the transmission or subtransmission voltage level. Much of the available renewable resources exist in locations where transmission is not available to transport the energy to the loads. In this case, new transmission lines are needed to make large quantities of renewable power available where it is greatly needed. The 2006 National Electric Transmission Congestion Study identified critical congestion areas based on current and projected growth. There are many barriers to get the needed transmission sited and installed including regulatory, cost recovery, and technical issues. Giving FERC exclusive jurisdiction in siting new transmission for renewables does not solve the problem about who pays for new transmission. A technical challenge that has not been addressed with the high penetration of renewables is grid stability. Utilities are now running into problems with maintaining frequency stability with large amounts of intermittent generation such as wind and solar. This is primarily due to the fact that currently deployed versions of these renewable resources are not dispatchable, in other words we only have wind energy when the wind blows and photovoltaic energy when the sun shines. This is unlike fossil, nuclear, and geothermal generation that we can dispatch as needed because we have control of the fuel source at the generation plant as needed. Utilities have compensated by running dispatchable fossil generators as “spinning reserve” to compensate for renewable intermittency, which is not always cost effective. There is concern that increasing the penetration of renewables on the electric grid will increase stability problems. A key technology that will enable the dispatchability of renewables and help alleviate the stability problem is energy storage. Significant additional research is needed in this area. Concentrating solar power is one renewable technology that can incorporate storage technology through molten salt.

The second approach to meeting the nations growing electricity demand is through the use of more distributed generation. This includes rooftop photovoltaic panels, photovoltaic arrays at distribution voltages, distributed wind, fuel cells, and other small generation resources that can be placed at or very near the load. This approach is attractive for remote load sites and where new transmission is not a viable option. There are numerous challenges that have been identified by increasing the penetration of renewables at the distribution level through the DOE Renewable Systems Interconnection (RSI) initiative. This set of studies completed in December 2007 outline the challenges for distributed photovoltaics in the categories of 1) Distributed PV System Technology Development, 2)Advanced Distribution Systems, 3)System Level Test and Demonstration, 4)Distributed Renewable Energy System Analysis, 4) Solar Resource Assessment, 6)Codes, Standards, and Regulatory Implementation. However, these studies only identify some of the problems and funding is not in place to address the identified challenges. While these studies focus on distributed photovoltaics, the concepts are applicable to most renewables.

In summary, transmission siting is just one element of the renewable interconnection challenge. Further research in renewable systems interconnection is needed to address the technology needs, codes and standards, business models, and regulatory issues to assure that we maintain a secure and reliable electric grid as we increase the penetration of renewables in the US.

ITC

Question 6. One of the most important issues facing the solar industry today is that the tax credits we passed in a bipartisan manner are set to expire. We must enact a long-term ITC extension as soon as possible. However, for the first time in the renewable tax credit history, the House Majority is insisting that the tax credits be "offset" by tax increases on other industries.

Many of you have submitted testimony highlighting the tremendous economic boost the ITC provides the solar industry. According to Dr. Marker from SHOTT, solar capacity additions in 2007 contributed $2 billion to the U.S. economy, creating 6,000 new jobs. And, it's forecasted that if the ITC is extended, 62,000 manufacturing and distribution jobs will be created.

Do you agree that these tax credits "pay for themselves" and therefore don't need to be paid for by raising taxes on other industries? Is the renewable industry concerned that this new "pay for" requirement can set a troubling precedent in that offsets will be required each time the existing tax credits expire.

Answer. We do not have the data to document the net "value" of the Investment Tax Credit. However, analysis for states in the southwest has shown that the development of CSP projects has a positive value to state and local economies over the lifetimes of the projects. Specific examples have been sited in several reports.[9 10 11]

Solar Resource Potential

Question 7. We currently have just over 400 MW of CSP installed capacity in this country, at a rate of about 16 cents per kilowatt-hour. A recent study suggests that solar energy could grow to 10% of the nation's power by 2025. Do you agree with that assessment? If so, what percentage will CSP contribute as opposed to Photovoltaic? Also, how long will it take to get solar power costs on parity with conventional power sources?

Answer. With the sustained incentive of the 30% ITC through 2017, the study results presented at the hearing predicts a deployment of 22 GW by 2025.[12] If the ITC were extended through 2025, the total deployment could be 40 GW as predicted by the same methodology presented in this reference. To reach 10% of the current U. S. grid capacity or about 100 GW by 2025 would require more aggressive incentives, significant streamlining of approval processes, and aggressive expansion of transmission capabilities.

CSP and photovoltaics are primarily focused on different market sectors. CSP is focused on the wholesale, utility-scale power market and photovoltaics are currently being applied to the "higher value" distributed, retail power market. The very large CSP installations are centrally sited, permitted, and maintained, but compete at utility generation rates. The bulk of current PV installations are distributed, either at the rooftop level or in relatively small power plants, are sited, permitted, and maintained by the owners, and compete financially on the customer side of the meter (retail pricing). The deployment scenarios and incentives are very different to reflect the needs and unique characteristics of these two markets. U. S. deployment of photovoltaics is predicted to be 20 to 25 GW by 2025 in the photovoltaics roadmap.[13] If the cost of photovoltaics falls as some think it may, it is possible that in the future PV may also compete in the utility-scale markets, as it currently does in Spain.

Even with the increase in the cost of commodities, CSP costs are projected to continue to decrease. The current cost of conventional pulverized coal power continues to increase with anecdotal costs indicated to be in the 5 to 7 cents/kWhr range. More importantly, uncertainty over future carbon regulation is resulting in the cancellation of orders for new pulverized coal plants. Carbon capture is projected to increase the cost of coal by an additional 5.4 cents/kwhr,[14] perhaps more if gasification technology is utilized. This means that the gap between the cost of generating a kWhr of electricity using pulverized coal and CSP and other sources of renewable electricity is growing smaller. As discussed during the testimony, once a solar plant is installed, the cost of electricity generated is relatively stable over the life of the plant, while conventional technology energy costs are highly dependent upon the cost of fuel.

Responses of Alex Marker to Questions from Senator Domenici

Question 1. You testified that eventually, solar storage could produce electricity 24/7 without the need for natural gas back-up. Currently, our solar storage capabilities are about 6 hours. When do you envision a 24/7 CSP plant?

Answer. It is technically possible to continuously operate a CSP trough plant based on technology available today. However it may not be economically competitive with baseload power generation. Many utility companies have a need for cost competitive peak power, which is what's driving demand for CSP technology.

CSP power plants, with thermal storage on the order of six hours, will provide utilities peaking capability that extends into the evening hours. With proper Federal incentives and policy, CSP can be competitive with other technologies for providing peak power, with the ultimate goal of continuous cost-competitive power generation. A long-term extension of the ITC will spur this research and will drive innovation.

Question 2. When your manufacturing plant comes on-line next Spring, how many megawatts of solar collectors will you be able to produce on an annual basis?

Answer. SCHOTT Solar anticipates having several production lines to manufacture receivers for parabolic trough concentrated solar power (CSP) plants. Each one of these lines will produce enough receivers capable of producing between 100MW and 200MW of annual power. Initial plans call for the facility to have two receiver production lines.

The broad range in output from each line is variable due to where the receivers are installed, and the overall size and efficiency of the projects they will be integrated into.

The facility is being designed with future expansion in mind, to accommodate the anticipated rapid growth of the utility-scale CSP market. SCHOTT Solar's further growth in Albuquerque is contingent upon worldwide market demand.

Transmission

Question 3. In the 2005 Energy Policy Act, Congress sought to address the critical issue of transmission siting through the National Interest Electric Transmission Corridor process. Even though these provisions haven't been fully implemented, and no line has been sited pursuant to EPAct, the NIETC process has proven controversial. Still, everyone here today has highlighted the critical need to bring more transmission on line to transport these renewable resources to load. Just this past weekend, the Albuquerque Journal ran an Op-Ed criticizing environmental groups—who want the "green" power but not the infrastructure that goes with it—for opposing needed transmission lines.

What more should Congress do in this important area? Some have called for Congress to provide FERC with exclusive jurisdiction to site new transmission for a renewable project. Please comment.

Answer. An expanded transmission grid is absolutely essential for moving renewable energy to customers. The current transmission grid was built decades ago to connect traditional fossil fuel generation with the load centers. Any carbon-free energy paradigm will require new transmission lines to once again connect customers with their desired power source—this time, solar, wind and other clean energy resources.

The lack of investment in our transmission grid will limit how much renewable energy can be delivered, and therefore developed. As you heard at the committee's June 17 hearing on transmission challenges, there is a clear need to deal with the problems around building new transmission lines.

One solution might be to give FERC exclusive jurisdiction to site new transmission lines, similar to its authority to site natural gas pipelines. Another solution may be multi-state regional cooperation, where public utility commissions, utilities and project developers could pool resources and expertise to propose new transmission lines to the benefit of all involved.

There are currently several cooperative efforts going on in the West—the Western Governors' Association Western Renewable Energy Zone initiative and California's Renewable Energy Transmission Initiative—which seek to both identify areas rich in renewable resources and the transmission necessary to move that power to end-use customers.

ITC

Question 4. One of the most important issues facing the solar industry today is that the tax credits we passed in a bipartisan manner are set to expire. We must enact a long-term ITC extension as soon as possible. However, for the first time in the renewable tax credit history, the House Majority is insisting that the tax credits be "offset" by tax increases on other industries.

Many of you have submitted testimony highlighting the tremendous economic boost the ITC provides the solar industry. According to Dr. Marker from SHOTT, solar capacity additions in 2007 contributed $2 billion to the U.S. economy, creating 6,000 new jobs. And, it's forecasted that if the ITC is extended, 62,000 manufacturing and distribution jobs will be created.

Do you agree that these tax credits "pay for themselves" and therefore don't need to be paid for by raising taxes on other industries? Is the renewable industry concerned that this new "pay for" requirement can set a troubling precedent in that offsets will be required each time the existing tax credits expire?

Answer. SCHOTT Solar ("the company") strongly believes that a long-term (8-year) extension of the ITC is an essential component in developing a long-term, sustainable market for solar energy in the United States. The company has not engaged in analyzing the financial impact of an extension of an ITC beyond the effects it will have on SCHOTT Solar's immediate business.

A study recently prepared by GE Energy Financial Services (June 2008) concluded that the ITC would " pay for itself" through continued investment and employment numbers.

It should be considered, and cautioned, that although the ITC is one of the most important factors in developing the solar market in the United States, it is not the only factor. Market development may not grow as forecasted.

Regarding setting a precedent for renewing the ITC—the market will rapidly develop, and consequently change, between now and when the ITC (if extended for the long-term) will be set to expire. SCHOTT Solar trusts in the Federal government to make an informed, and appropriate decision relating to extensions of the ITC, based on market analysis available at that time.

Solar Resource Potential

Question 5. We currently have just over 400 MW of CSP installed capacity in this country, at a rate of about 16 cents per kilowatt-hour. A recent study suggests that solar energy could grow to 10% of the nation's power by 2025. Do you agree with that assessment? If so, what percentage will CSP contribute as opposed to Photovoltaic? Also, how long will it take to get solar power costs on parity with conventional power sources?

Answer. The potential of solar energy is almost limitless, especially in the Southwest of the United States, where, if just 2.5% of the usable land were converted into CSP power plants, it would satisfy the entire nation's energy needs (at 2006 usage).

Before directly answering the question, it should be noted that the 16 cents per kilowatt-hour is an aggregated figured based upon the "first generation" of parabolic trough CSP plants (SEGS). Costs will continue to decrease through technological developments, economies of scale in constructing the sites, and improvements in transmission. At Acciona Energy's "Nevada Solar One " facility in Boulder City, NV, it is estimated that the cost to generate electricity from the facility is fixed at 3 cents per kilowatt hour (not inclusive of amortizing the cost of constructing the facility).

Energy Mix

Whether or not the nation will be able to consume 10% of its energy from solar powered sources, is contingent upon the speed and ease in which large-scale solar facilities will be allowed to set-up, operate, and provide power to the grid. A long-term extension of the ITC is a key component in allowing this rapid market development. A Federal Renewable Portfolio Standard (RPS) would motivate utilities to buy and consume renewable energy. Streamlining and standardizing grid connection and transmission will also create a powerful climate for the rapid growth of CSP.

Percentage of CSP vs. PV

CSP and PV are complementary technologies (much like wind and solar are complementary). PV is mainly used for distributive power generation, where the power produced is consumed at that location. With only a handful of notable exceptions, PV installations generally do not exceed 1MW in capacity.

CSP, however, is a utility-scale, central station technology where installations generally achieve production exceeding 50MW capacity. Although it is possible to construct and operate smaller scale CSP power plants, through economies of scale, they are not cost-competitive.

Due to the sheer size of CSP power plants, as well as the speed in which they're able to be constructed (for reference, Accciona Energy's Nevada Solar One was producing power a year after ground was first broken), and no foreseen raw material supply constraints, under the right climate—tremendous CSP growth is possible. As a point of reference, there are currently 4, 000 MW of CSP projects currently in the planning / permitting stage. With a clear signal of support from the Federal government in this clean, fixed cost, domestically produced energy source—many of these projects will shift from planning to construction and many more will enter the planning stage.

It is difficult to assign an exact figure for PV growth over the eight year term of an ITC extension, as the nature of projects are generally much smaller. One can look at PV manufacturing capacity (which is rapidly developing) but it is unknown how much PV will be imported from overseas, and conversely, how much will be exported from the U.S. It is for this reason, that I apologize I am unable to provide the committee with an exact answer to this question.

Closing Remark

Distinguished committee members, on behalf of the solar industry, and SCHOTT Solar, we appreciate your continued efforts to focus on the promise and possibility of solar energy. The benefits of the technology are numerous, not only from an environmental standpoint, but also as an economic engine. Solar energy has tremendous support across the country. 94% of Americans, in a recent poll, stated it is important for the U.S. to develop and use solar energy, and approximately 75% of Americans favor a long-term extension of the ITC. This is an issue we all agree on.

The United States and New Mexico stand to reap a tremendous economic benefit from a long-term extension of the ITC. There is no reason why New Mexico can not take a global lead in renewable energy.

In twenty years I am confident that, when flying across the desert Southwest, a glimmer of light will catch my eye from a parabolic mirror, focusing the sun's energy on a receiver, which will be providing clean, reliable, fixed cost energy across the U.S. And I will think back to this time, and the hard work we are all undertaking.

It is through your leadership and guidance that the United States will return to its position as a clean energy leader.

Responses of Fong Wang to Questions from Senator Domenici

Question 1. You testified that last year, approximately 12% of PG&E's electric delivery mix was comprised of renewable resources. How much of this is comprised by solar? How much wind have you added to meet California's RPS requirement? Are you on target to meet the State's goal of 33% of renewable energy by 2022? Isn't there an initiative to raise this goal even further, to 50%?

Answer. In 2007, the breakout of PG&E's 12 percent California-eligible[15] renewable electricity delivery mix was as follows: solar, less than 1 percent; wind, 15 percent; California-eligible small hydroelectric, 21 percent; geothermal, 30 percent, and biomass and waste, 34 percent. (Note: the distributed solar installed by PG&E's customers does not count for PG&E's RPS requirement.)

Since the California RPS was created in 2002, PG&E has signed 39 contracts for 2,612 MW (up to 3,195 MW if all options are exercised). The amount of wind and other resources added are as follows:

	contracts	MW
Solar	6	1343
Wind	8	584
Geothermal	8	553
Bioenergy	14	129
Wave	1	2
Small Hydro	2	1
	39	2612

A 33 percent renewable energy standard has been proposed by the California Air Resources Board (CARB) as a key pillar to driving down greenhouse gas emissions in California. PG&E is reviewing that option and looking forward to working with the CARE and other interested parties going forward. We are committed to being a constructive part of California's drive to increase its use of renewable resources. We do believe, however, that as these requirements are debated and established, policy makers should carefully consider and must ensure that:

1. Appropriate tax and financial incentives are available,
2. Sufficient transmission has been built or approved to deliver newly developed renewable energy to customers,
3. Consensus is reached on the maximum price customers should be expected to pay for renewable energy, and
4. Reliability isn't jeopardized by over-reliance on intermittent energy resources.

We are confident these issues will be addressed, as they are necessary to ensure the viability and success of RPS requirements.

We do not support the current ballot proposal in California to raise the RPS level to 50%, and have joined a coalition with environmental groups, organized labor, and renewable energy producers to oppose this well-intended, but poorly drafted ballot initiative. We would recommend that you contact the coalition Californians Against Another Costly Energy Scheme (www.noprop7.com) for further information on this matter as it has been authorized to speak on behalf of all of its members.

Question 2. I understand that PG&E is working through California's Renewable Energy Transmission Initiative. Please elaborate. Are you working on any interstate transmission policies or are you concentrating wholly intrastate?

Answer. PG&E is an active participant in California's Renewable Energy Transmission Initiative, or "RETI." In addition to California, the footprint of the RETI analysis includes British Columbia, Washington, Oregon, Nevada, Arizona and Baja Mexico. This process will identify renewable resource areas that are well-suited for major new transmission investments, as well as specific transmission projects to access the identified areas. RETI is being developed in phases with the first phase focused on identifying the most promising renewable resource areas and the second phase focused on identifying conceptual transmission plans necessary to gain access to these areas. The initiative is modeled on the approach used in the Tehachapi area in California and also on the work done by ERGOT in Texas. The current scope of the first phase to identify renewable resource areas includes areas outside California. We recommend that federal legislation avoid creating conflicts with RETI and any other similar state-led initiatives in this area. Federal policies related to these efforts include interconnection policies that govern the process for interconnection studies and cost allocation for constructing any necessary transmission upgrades to support renewable development. Federal land use policies will also be a key issue in the ultimate development of the renewable resource areas and the associated transmission necessary to access those resources. Some of the lead federal agencies are participating in the RETI process. The extent to which the land use policies of these agencies align with the goals of RETI has not yet been specifically determined.

PG&E is also supporting the Western Electricity Coordinating Council Transmission Expansion Planning and Policy Committee workgroups that are identifying interstate transmission options for supporting various renewable energy development scenarios in the West.

Question 3. How are you dealing with the transmission necessary to transport your solar energy?

Answer. PG&E and other California-based investor owned utilities have executed agreements to purchase power from a number of solar energy projects, however, the identification of the transmission method of service for these projects have been delayed due to the large number of generation projects in the CAISO interconnection queue. PG&E is actively participating in an effort by the CAISO to address its generation interconnection process with the aim of providing a clear path to interconnecting renewable resources in California. PG&E and the CAISO have also identified a major transmission upgrade opportunity to address reliability needs, improve access to energy storage facilities and improve access to renewable resources in southern California. The project does not specifically require interconnection of solar energy resources but is part of transmission expansion plans to access and integrate renewable resources in California.

Question 4. In the 2005 Energy Policy Act, Congress sought to address the critical issue of transmission siting through the National Interest Electric Transmission Corridor process. Even though these provisions haven't been fully implemented, and no line has been sited pursuant to EPAct, the NIETC process has proven controversial. Still, everyone here today has highlighted the critical need to bring more transmission on line to transport these renewable resources to load. Just this past weekend, the Albuquerque Journal ran an Op-Ed criticizing environmental groups—who want the "green" power but not the infrastructure that goes with it—for opposing needed transmission lines.

What more should Congress do in this important area? Some have called for Congress to provide FERC with exclusive jurisdiction to site new transmission for a renewable project. Please comment.

Answer. In California, PG&E has strongly supported efforts at the California Public Utility Commission (CPUC) and the California Independent System Operator (CAISO) intended to provide California utilities with the tools they need to proactively plan and develop transmission facilities for renewable sources of electricity. These efforts include the development of a category of needed transmission projects under the CAISO tariff for renewable trunk lines, as well as "backstop" cost recovery proposals at the CPUC intended to encourage utilities to exercise their right under FERC interconnection policy to provide up-front funding for RPS-related network upgrades where doing so makes sense for customers and the state's RPS goals.

Planning for and expansion of transmission is appropriately performed by transmission owners working in conjunction with appropriate state or regional transmission system operators. The application of FERC backstop siting authority should be consistent with state and regional planning efforts to develop and maintain coordinated transmission planning processes.

The states are generally best suited to address local and state issues concerning the siting and design and environmental review of state projects. However, PG&E sees FERC's backstop siting authority as a potential useful mechanism for ensuring the efficient processing of transmission corridors when states have not acted.

Specifically, in cases where local interests may prevent transmission facilities from being sited and the state has withheld approval, FERC involvement can be useful in moving the process along. For example, FERC's backstop siting authority may be most appropriate for interstate corridors with significant congestion, and where proposed interstate projects require approvals by multiple states with potentially conflicting objectives.

Question 5. One of the most important issues facing the solar industry today is that the tax credits we passed in a bipartisan manner are set to expire. We must enact a long-term ITC extension as soon as possible. However, for the first time in the renewable tax credit history, the House Majority is insisting that the tax credits be "offset" by tax increases on other industries.

Many of you have submitted testimony highlighting the tremendous economic boost the ITC provides the solar industry. According to Dr. Marker from SHOTT, solar capacity additions in 2007 contributed $2 billion to the U.S. economy, creating 6,000 new jobs. And, it's forecasted that if the ITC is extended, 62,000 manufacturing and distribution jobs will be created.

Do you agree that these tax credits "pay for themselves" and therefore don't need to be paid for by raising taxes on other industries? Is the renewable industry concerned that this new "pay for" requirement can set a troubling precedent in that offsets will be required each time the existing tax credits expire?

Answer. PG&E supports establishing long-term tax credits, and the market certainty that provides, such as the 8-year extension proposed in recent federal legislation. This will enable the solar industry to obtain the continued financial investment necessary to see these multi-year projects through from design, to construction, and to operation. This long-term tax credit approach will also enable the industry to establish a sufficient manufacturing base to achieve the economies of scale necessary to reduce costs and ultimately minimize the need for such subsidies. Conversely, the potential of this industry will be delayed or not be realized in an environment characterized by year-to-year extensions, and the market uncertainty created by such annual legislative processes.

Question 6. We currently have just over 400 MW of CSP installed capacity in this country, at a rate of about 16 cents per kilowatt-hour. A recent study suggests that solar energy could grow to 10% of the nation's power by 2025. Do you agree with that assessment? If so, what percentage will CSP contribute as opposed to Photovoltaic? Also, how long will it take to get solar power costs on parity with conventional power sources?

Answer. The study cited, "Utility Solar Assessment (USA) Study, Reaching Ten Percent Solar by 2025", projects that solar PV would provide 8 percent and CSP would provide 2 percent of total U.S. electricity. It would require an installation of 255 GW of solar by 2025 at a Compound Annual Growth Rate of 33 percent for PV and 28% for CSP.

The amounts projected are achievable, but to reach those values by 2025 would require 100 percent of the many assumptions to happen perfectly and on schedule. Experience tells us that rarely does that happen. We would expect the actual process to take somewhat longer.

Achieving grid parity between solar and local utility rates will be affected by several factors, including, among others, utility rate design, solar insolation in a given area, increases in fossil fuel costs, and carbon costs. It is expected that grid parity will be reached prior to the 2025 reference date, possibly as early as 2015 in some areas.

End Notes

* Figures 1–4 have been retained in committee files.
** Figures 1–6 have been retained in committee files.
*** Presentation has been retained in committee files.
**** All figures have been retained in committee files.

1 Report available at: http://www.nrel.gov/csp/troughnet/pdfs/kearney_wga_overview.pdf.

2 Sergeant and Lundy LLC Consulting Group, "Assessment of Parabolic Trough and Power Tower Solar Technology Cost and Performance Forecasts", Chicago, IL: NREL/SL-5641, October 2003.

3 California CPUC Resolution E-4118 Adopting the 2007 MPR (Market Price Referent), 4 October 2007. Document available at: http://docs.cpuc.ca.gov/published/Final_resolution/73594.htm.

4 Department of Energy Solar Energy Technologies Program, Multi Year Program Plan, 2008-2012, April 2008. Plan available at: http://www1.eere.energy.gov/solar/pdfs/solarlprogramlmyppl2008-2012.pdf.

5 Frosch, Dan. "Citing Need for Assessment, U.S. Freezes Solar Energy Projects," New York Times, 27 June 2008. http://www.nytimes.com/2008/06/27/us/27solar.html

6 As defined in Senate Bill 1078, which created California's renewable portfolio standard, an eligible renewable resource includes geothermal facilities, hydroelectric facilities with a capacity rating of 30 MW or less, biomass, selected municipal solid waste facilities, solar facilities and wind facilities.

7 The fourth contract is with San Joaquin Solar for 106.8 MW in Fresno County, CA.

8 Assessment of Parabolic Trough and Power Tower Solar Technology Cost and Performance Forecasts, prepared for the Department of Energy and National Renewable Energy Laboratory, SL-5641, May 2003.

9 The Economic Impact of Concentrating Solar Power in New Mexico, The University of New Mexico, Bureau of Business and Economic Research, December 2004.

10 The Potential Economic Impact of Constructing and Operating Solar Power Generation Facilities in Nevada, Final Report, R. Keith Schwer and Mary Riddel, Center for Business and Economic Research, University of Nevada, Las Vegas, July 2003

11 Economic, Energy, and Environmental Benefits of Concentrating Solar Power in California, Deliverable 3 Final Report, Black and Veatch, Prepared for National Renewable Energy Laboratory Under Subcontract AEK-5-55036, September 2005

12 Department of Energy Solar Energy Technologies Program, Multi Year Program Plan, 2008-2012, April 2008. Plan available at: http://www1.eere.energy.gov/solar/pdfs/solarlprogramlmyppl2008-2012.pdf.

13 Solar Electric Power, The U.S. Photovoltaic Industry Roadmap, May 2001, available at http://photovoltaics.sandia.gov/docs/PDF/PVlRoadlMap.pdf

14 Reducing U. S. Greenhouse Gas Emissions: How Much at What Cost?, U. S. Greenhouse Gas Mapping Initiative, Executive Report, December 2007. (aka, the McKinsey Report)

15 As defined in California Senate Bill 1078, which created California's renewable portfolio standard, an eligible renewable resource includes geothermal facilities, hydroelectric facilities with a capacity rating of 30 MW or less, biomass, selected municipal solid waste facilities, solar facilities and wind facilities.

In: Solar America: How, What and When?
Editor: Nash M. Perales

ISBN: 978-1-60741-333-2

Chapter 5

UTILITY-SCALE SOLAR POWER: OPPORTUNITIES AND OBSTACLES: FIELD HEARING BEFORE THE SUBCOMMITTEE ON ENERGY AND ENVIRONMENT, COMMITTEE ON SCIENCE AND TECHNOLOGY, HOUSE OF REPRESENTATIVES, ONE HUNDRED TENTH CONGRESS

U.S. Government Printing Office

PURPOSE

On Monday, March 17, 2008 the House Committee on Science & Technology, Subcommittee on Energy and Environment will hold a hearing entitled, *''Utility-Scale Solar Power: Opportunities and Obstacles,''* at the Pima County Administration Building Hearing Room, Tucson, Arizona.

The Subcommittee's hearing will explore the potential for utility-scale solar power to provide a significant fraction of U.S. electric generating capacity and the challenges to achieving this goal. The specific technologies to be discussed include solar thermal technology, concentrating photovoltaics and distributed solar power. Transmission, regulatory and financial issues will also be examined, along with a look at the government and private industry roles in the development of utility-scale solar power—and enabling productive partnerships between them.

WITNESSES

- **Mr. Mark Mehos** is the Program Manager for the Concentrating Solar Power Program at the National Renewable Energy Laboratory. Mr. Mehos will provide an overall assessment of the available resource size for solar energy in the U.S. and an introduction to the known technologies that may take advantage of solar power on a large scale.

- **Mr. Tom Hansen** is the Vice President of Environmental Services, Conservation and Renewable Energy at Tucson Electric Power. Mr. Hansen will describe a ''Solar Grand Plan'' to provide more than half of the U.S.'s electricity from solar power by 2050.
- **Ms. Kate Maracas** is the Vice President of Arizona operations at Abengoa Solar. Ms. Maracas will describe the current state of solar thermal technology and the near- and long-term economic costs and benefits of large-scale solar power in general.
- **Ms. Valerie Rauluk** is the Founder and CEO of Venture Catalyst, Inc. Ms. Rauluk will describe the current state of distributed and concentrating photovoltaics and provide an assessment of how the marketplace for solar energy will change over the next 10 years.
- **Ms. Barbara Lockwood** is the Manager of Renewable Energy for Arizona Public Service. Ms. Lockwood will provide the perspective of utilities on the ability for large-scale solar power to be a significant competitor in the U.S. energy sector over the next 50 years.
- **Mr. Joe Kastner** is the Vice President of Implementation and Operations for MMA Renewable Ventures LLC. Mr. Kastner will describe his company's experience with installing and managing the Nellis Air Force Base solar array and ways to enable productive partnerships between government and renewable energy industries in general.

BACKGROUND

An article in the January 2008 issue of *Scientific American* titled ''A Solar Grand Plan'' outlined a potential path to providing nearly 70 percent of U.S. electricity demand and 35 percent of its total energy demand, including transportation, with solar power by 2050. It is also estimated that if fully implemented, the plan would reduce U.S. carbon dioxide emissions to 62 percent below 2005 levels. Approximately $420 billion in various government subsidies from 2011 to 2050 would be required to fund the necessary infrastructure and make solar power cost-competitive.

Several types of technology would be needed to follow through on such a plan. Photovoltaics (PV), which convert sunlight directly to electricity, are the most familiar. Vast arrays of PV cells can be deployed in the Southwest covering multiple square miles to generate hundreds of megawatts of electricity per field. A variation on this technology, known as concentrating photovoltaics (CPV), uses lenses or mirrors to concentrate sunlight onto high-efficiency solar cells. These solar cells are typically more expensive than conventional cells used for flat-plate PV systems. However, the concentration decreases the required cell area while also increasing the cell efficiency. PV and CPV systems may employ any of a number of electrical energy storage technologies for use during periods of passing clouds or into the evening.

An alternate technology could also be used. Solar thermal technology produces electric power by converting the sun's energy into high-temperature heat with various mirror configurations. The heat is then used to power a conventional generator. Solar thermal plants consist of two parts: one that collects solar energy and converts it to heat and another that converts heat energy to electricity. Just as batteries may assist PV systems, molten salts and other forms of thermal energy storage technology allow this heat to be retained for later use in generating electric power.

An expansive new transmission and distribution system would be required for the remainder of the country to take full advantage of the immense solar resource in the American Southwest. A 2005 study commissioned by the Western Governors' Association estimated that solar energy from the Southwest alone could provide up to 6,800 GW of electricity to the U.S. To put this in perspective, the electric generating capacity of the entire country is currently about 1,000 GW. However, the existing system of alternating-current (AC) power lines would lose a significant fraction of its energy over long hauls so the Solar Grand Plan recommends building a new backbone of high-voltage, direct current (HVDC) power transmission lines and coupling this to sites near population centers that may utilize another form of electrical energy storage technology known as compressed air energy storage (CAES).

Under this scheme, electricity generated from solar power plants hundreds of miles away compresses air and pumps it into vacant underground caverns, abandoned mines, aquifers and depleted natural gas wells. The pressurized air is released on demand to turn a turbine that generates electricity, aided by burning small amounts of natural gas. Citing a study by the Electric Power Research Institute (EPRI) and the natural gas industry, the Plan affirms that suitable geologic formations exist in 75 percent of the country often close to metropolitan areas, and that a national CAES system would look similar to the current U.S. natural gas storage system.

The Plan assumes relatively small increases in PV solar-to-electric efficiency from 10 percent today to 14 percent in 2050 and increases in efficiency for solar thermal technology from 13 to 17 percent. The Plan also assumes significant reductions in installed cost and electricity price based on economies of scale reaching 5–9 cents/kWh. (Today's rates for these systems in the U.S. are 16–18 cents/kWh and average overall electricity rates are currently 5–15 cents/kWh depending on the region.)

Though not directly addressed by the Solar Grand Plan, one other method to generate solar power on a large scale is distributed generation (DG) which consists of smaller facilities on otherwise unused real estate (roof-tops and sites of 10 to 500 acres) located near the load demand and dispersed throughout many communities. DG systems typically produce under 20MW of power and may consist of PV and CPV components. By providing power near or directly at the point of use, DG may offer a more cost-effective near-term solution in many areas of the country.

Ms. Giffords. This hearing will come to order. Good morning, everyone. It is my great, great honor and privilege to welcome you all this morning to a field hearing of the Subcommittee on Energy and Environment entitled ''Utility-Scale Solar Power: Opportunities and Obstacles.''

I want to welcome everyone here to Tucson. I want to thank Chairman Richard Elý′as for having us here in the Supervisor's Headquarters. Richard, where are you? Thank you so much for having us here today.

We also have with us Councilwoman Nina Trasoff. Thank you for all the work that you are doing, one of our local elected officials, I appreciate having all your support.

Rarely does a meeting in Congress go by when I do not have an opportunity to talk to my colleagues about how extraordinary southern Arizona is, and so that is one of the reasons why I am so pleased to have so many of our colleagues here with us today to enjoy this extraordinary part of the world that we call home.

Many thanks to my colleagues who are interested in utility-scale solar power for coming. All have given up time in their district work periods to be here in southern Arizona. We are honored to have Members from around the country with us today. I believe it is a testament to the high degree of energy and interest that we have in solar technology.

In particular, I would like to extend a very special welcome to the two most senior Members of the Science and Technology Committee. The Chairman of the Full Committee, Mr. Bart Gordon from Tennessee, unfortunately, was detained in Washington due to weather and mechanical problems with his airplane, but he is going to be joining us by telephone in just a few minutes.

But, I would like to thank Ranking Member, Mr. Ralph Hall from Texas. Ralph Hall has been on the Science and Technology Committee for 28 years, and also Energy and Commerce. So, we are very pleased to have him with us today.

Also with us is Representative Dan Lipinski of Illinois, Vice Chair of the Full Committee, Representative Jim Matheson from the State of Utah, thank you for coming, and Representative Harry Mitchell, a fellow Member of the Arizona Congressional Delegation, from Tempe, Arizona.

No one can remember the last time that we had so many Members of Congress in southern Arizona for an actual field hearing. So, I am particularly pleased that we are all here on a topic as important as solar energy.

I would like to extend a very warm welcome to our witnesses who are here today. We are glad that you are here to share in your expertise, to enlighten the Committee and members of the public about the experiences that you have and the thoughts that you have. It was challenging to only have six panelists when we have so many talented people that are experts in solar technology, but we are very pleased that you are here to join with us today.

We also have many smart and talented members of my Solar Advisory Committee that are in the audience today, and I want to thank you as well for taking this part of the world and making it so focused on solar energy and the possibilities that lay ahead of us.

Finally, a welcome to all the members as well, just of the general public, who are here because they care about what is happening with the future of energy technology, and again, I want to thank you, our Members today, but also watching live on the Internet, and we know we have a streaming video and we will capture that for people who are not able to participate.

In the Science and Technology Committee, it is common for our Chairman, Mr. Gordon, to refer to us as the Committee of big ideas. The notion is certainly well grounded from the history of the Committee. Just last week we celebrated 50 years of the Science and Technology Committee, with Bill Gates coming to talk to us. It is a Committee that oversaw the days of NASA and winning the space race, the wake of Sputnik, but the Chairman's statement is as much, I believe, about the future as it is about the past. It expresses a belief which I share, that the greatest days of American innovation are the days which lay ahead. So, in my view, it could not be more fitting for the Committee to turn its attention to solar power.

Solar power is a big idea, whose time has come. And, like the space program, solar is an idea that can shape our nation in significant and positive ways. In the coming months, in the coming years, we will face critical decisions on how to address climate change, reduce our dependence on foreign energy, and boost our economic competitiveness.

The beauty of solar power is that it offers an elegant solution to all three of these challenges. Imagine what it would be like if every time that it rained it rained oil, big black drops falling from the sky. Don't you think that we would find some way to run around with a big bucket and collect all of that energy that was falling from the sky? I know this sounds like an absurd picture, but the reality is that what we have outside today is something very comparable to that. Literally, we have useful energy pouring out of the sky, and nowhere does it rain sunshine with greater intensity and consistency than in the American southwest.

In fact, some studies show that there is enough sunshine in the southwest to power almost our entire nation. One of these studies was recently covered on—actually, was brought forward on the front cover of the *''Scientific American Magazine.''* So, in other words, the southwest is home to a national treasure that streams from the sky almost every day.

That sounds like a good enough reason to start developing an effective solar bucket, and while we are at it, let us make it a really big bucket.

The focus of this hearing is not just about any kind of solar power, it is about utility-scale solar power. Utility-scale refers to large installations that can generate significant amounts of electricity for the grid, but with free fuel and, virtually, no pollution. Developing solar installations on this scale creates unique opportunities, but it also presents unique challenges.

So, we look forward today to hearing from our witnesses about both. Our goals in holding this hearing are to explore five key issues. First, the potential scale of solar power in America. Second, the current state of technology. Third, the benefits to our nation of embracing this energy source. Fourth, the obstacles to developing solar power in a big way. And finally, the policies that can help us overcome these obstacles.

The time for solar is now. Technologies are proving, the costs are falling, and the reasons to adopt it are compelling. We need to truly understand the potential of this energy source and how we can unleash it. So, that is what today's hearing is all about.

So, we should get started. Since we do not have Congressional hearings in Tucson every day, I want to briefly explain how we are going to proceed. First, some of my colleagues will make opening statements. Then we will have a chance to hear from each of the witnesses in turn. We ask our witnesses, because of our time consideration, to keep their testimony to five minutes, and we have our technology here on the table to indicate when your time is being close to up.

Then, following the witnesses, each Member will have five minutes to ask questions. Once all the Members have asked questions, if time remains we will have a chance to recycle and go back to the beginning.

I know that some of my colleagues have planes to catch, so we are planning on wrapping up around 2:30.

Now, following the conclusion of the formal hearing, I would like to take a short ten-minute break. I encourage people who are interested in the community to stay around, because we are going to ask our panelists to come on the dias and be able to directly answer your questions.

With that, I would now like to yield to Mr. Hall for his opening statement.

[The prepared statement of Acting Chair Giffords follows:]

PREPARED STATEMENT OF ACTING CHAIR GABRIELLE GIFFORDS

Good morning. It is my great privilege today to convene this field hearing of the Subcommittee on Energy & Environment, entitled *''Utility-Scale Solar Power: Opportunities and Obstacles.''* I want to welcome everyone to Tucson.

Rarely does a hearing go by where I do not talk about Arizona so you can imagine what a pleasure it is to be able to show my fellow Committee Members why I am so proud of our community and the work we are doing together on Solar energy.

Many thanks to my colleagues from the Science and Technology Committee. They have all given part of their District work period to come to southern Arizona today.

We are honored to have with us Members from all over the country. This is a testament to the high level of interest in solar power, and to its relevance to the whole Nation.

In particular, I would like to extend a very special welcome to the two most senior Members of the Full Science and Technology Committee:

- The Chairman of the Full Committee, Mr. Bart Gordon, of Tennessee, who was unfortunately detained in Washington due to weather and mechanical problems with his airplane. He should be joining us by phone in just a bit. And
- The Ranking Member, Mr. Ralph Hall, of Texas.

Thank you both for coming.

Also with us today are:

- Rep. Dan Lipinski of Illinois, Vice Chair of the Full Committee,
- Rep. Jim Matheson from Utah, and
- Rep. Harry Mitchell, a fellow Member of the Arizona delegation.

No one can remember the last time that so many Members of Congress came together in Tucson for a Field Hearing. I am particularly pleased that we are here on such an important topic.

Thank you all for making the special effort to be here today.

I would like to extend a very warm welcome to our witnesses. We are glad you are here to share your expertise with the Committee. It was an incredible challenge to narrow our panel down to just six.

There are so many smart and talented people with important perspectives on these issues, including many members of my own Solar Advisory Council.

I wish we could fit them all at the witness table, but space and time constraints prevent us from doing so.

I thank all of these people for their important contributions to solar power and their work with my office on our solar initiatives. We value their expertise, and I will continue to seek their counsel and collaboration as we move forward.

Finally, a special welcome to all the members of the community who are here today. Thank you for your interest in this critical issue.

In the Science and Technology Committee it is common for our Chairman, Mr. Gordon, to refer to us as the ‘‘Committee of Big Ideas.’’

This notion is certainly well-grounded in history. Formed in the wake of Sputnik and initially charged with winning the space race, the Committee is now celebrating 50 years of promoting big ideas in American science and technology.

But the Chairman's statement is as much about the future as it is about the past. It expresses the belief—which I share—that the greatest days of American innovation lie ahead of us.

So in my view it could not be more fitting that the Committee is turning its attention to solar power. Solar is a BIG IDEA whose time has come.

And like the space program, solar is an idea that can shape our nation in significant and positive ways.

In the coming months and years, we will face critical decisions on how to address climate change, reduce our dependence on foreign energy, and boost our economic competitiveness.

The beauty of solar power is that it offers an elegant solution to all three of these pressing concerns.

Imagine what it would be like if every time it rained, it rained oil—big, black drops falling from the sky. Don't you think we'd find a way to catch some of that bounty from the heavens? I think we'd be running around with big buckets, scooping up every available drop.

As absurd as that picture may be, with solar energy we have something just as good—useful energy that is literally pouring down from the sky.

And nowhere does it ‘‘rain’’ sunshine with greater intensity and consistency than in the American Southwest. In fact, some studies show there's enough sunshine in the Southwest to power almost our entire country! One of these studies was recently reported in a cover story in *Scientific American.*

In other words, the Southwest is home to a national treasure that streams from the sky almost every day. That sounds like a good reason to get serious about developing an effective solar bucket.

And while we're at it, let's make it a big bucket. The focus of this hearing is not just any kind of solar power, it is utility-scale solar power.

Utility-scale refers to large installations that can generate significant amounts of electricity for the grid, but with free fuel and no pollution.

Developing solar installations on this scale creates unique opportunities, but it also has unique challenges. We look forward to hearing from our witnesses today about both.

Our goals in holding this hearing are to explore five issues:

- the potential *scale* of solar power in America
- the current state of solar *technology*
- the *benefits* to our nation of embracing this energy source
- the *obstacles* to developing solar power in a big way, and
- the *policies* that can help us overcome those obstacles.

The time for solar is now: technologies are improving, costs are falling, and the reasons to adopt it are increasingly compelling.

We need to truly understand the potential of this energy source, and how we can unleash it. That's what today's hearing is about.

Mr. Hall. Thank you very much, Madam Chair, and I am honored to be here.

What she did not tell you is that I am the oldest guy in the United States House of Representatives, and that makes me the dean. I am 84 years old, but I was running at 5:45 this morning, two to three miles every morning. I am trying to stay young to keep up with my grandchildren. I get a little sick of some of them telling old man jokes, Madam Chair, about me. The latest one was that a woman's husband was about to quit golfing because he was 92, and he could not see where the ball was going. And, she hated to see him leave golf and be at home all the time. She said, well, my brother is 94 and he does not golf but he likes exercise. He has wonderful vision and he can see, I will bet he could tell you where your ball went. And, they worked that out. That following Monday out on the golf course he hit that ball, and it was way up in the air. He said, Orville, are you watching it? He said, yeah, I am watching it. He said, is it still up there? He said, yeah. Can you see it? Yeah, I can see it. Has it hit yet? He said, yeah, just hit. He said, where did it go. And, he said, I cannot remember.

So, us senior Members have problems of all kinds. But in this campaign—I just won the primary election, and the *''Dallas News''* called me an old geezer. And I made the argument, Madam Chairman, that it does not hurt to have—I do not recommend a whole floor of old 84-year-old guys or women, but it does not hurt to have one old geezer up there. And, I had all kind of call-ins and letters and everything, and finally the guy that won the contest told me my motto should be, ''Win One for the Geezer," and that is what we did.

But, I am glad and honored to be here with you in Tucson, and this very important hearing on solar energy. I am anxious to listen. I am here more to listen than I am to talk. The longest speech I will make is one that I will be reading to you right now in a few minutes.

I just want to say that I have often said that our country needs to become more energy independent, and to do that we need to use all the forms of energy. Americans, we have many forms of energy, and we just passed an energy bill a year and a half ago that had some incentive for every form of energy. And, energy is important, it is important in that if we solve our energy problems we also solve our war problems, because energy or lack of energy is the cause of most wars. It is the causation of it.

Japan did not dislike this country. Cordell Hull and Henry Stimson had cut their oil off. They had 13 months of national existence with no oil. So, we had to know they were going to break out and go south into Malaysia or somewhere. That was an energy war, it was not because they did not like us. Japan today is probably the best friend we have in the world, the best partner we have, I think. And, I would like to see them arm again, because they know how to deal with the Chinas, and the Koreas, and all that, but that is not what we are here about today. We are here about energy, and, of course, when George Bush, the elder George Bush, sent 450,000 youngsters to Iraq ten or 11 years ago, that was to keep a bad guy named Saddam Hussein from getting his foot on half the known energy in that part of the world. That was an energy war, in my opinion.

And, that is how important this hearing is, and solar is such an important part of energy. A balanced solution can reduce our dependence on foreign sources of energy, and that is one thing we really need to do, to make our air and water cleaner, most importantly, a viable solution reduces the cost of energy for Americans so that their economy can continue to grow.

As demand rises by an estimated 40 percent in the electricity sector by 2030, we are going to need solutions that keep America economically prosperous and competitive. As is evident here in the desert of the American southwest, the sun provides an abundant source of energy. Citizens have been using this source for years to power many small and some large-scale projects and devices. And, an entire small community of astronauts live on the International Space Station that is run entirely on solar energy. Yet, solar energy makes up a very small proportion of our overall consumption picture, only 0.4 percent of global energy demand is met by geothermal solar and wind energy combined.

In the United States, statistics by the Department of Energy from 2004 indicate that solar energy accounted for one percent of the total U.S. energy consumption, and 0.2 percent of our electricity generation.

Part of the reason this resource is not widely used is that sunlight is not constant, it is not focused. The amount of energy generated depends on the time of day, location, and weather conditions. In order to use solar energy for stable grid operations, we need better storage techniques. To this end, I introduced a bill last year, the *Energy for America Act,* which included the provision permitting research and development and demonstration of energy storage technologies for electricity transmission and distribution. I am pleased that this provision was included in the energy bill signed into law last year.

So, Madam Chair, I look forward to hearing this testimony today by this very esteemed panel of individuals, on how America can better harness the power of the sun to generate more utility-scale power that lowers the cost to taxpayers and consumers.

We write law up there. We write the law, but we write it based on people that know more about what we are talking about than we do, and that is this panel and people just like you. We will take your testimony, the rest of this committee, and the rest of Congress will have the opportunity to read it. Some of them will read it, but it can be the major part of a bill that would be introduced later, probably by Madam Chair with many of us being her co-sponsors.

With that, I thank you, and I yield back my time.

[The prepared statement of Mr. Hall follows:]

PREPARED STATEMENT OF REPRESENTATIVE RALPH M. HALL

Thank you. I am pleased to be here today in Tucson, Arizona for this important hearing on solar energy and I want to thank Rep. Giffords for organizing this gathering.

I have often said that America needs to become more energy independent, and to do that we need to use *all* forms of energy. Americans currently have, and will continue to need, reliable and affordable domestic energy. Citizens are rightfully concerned about rising energy prices and protecting the environment. A balanced solution reduces our dependence on foreign sources of energy while also making our air and water cleaner. Most importantly, a viable solution *reduces* the cost of energy for Americans so that our economy can continue to grow. As demand rises by an estimated 40 percent in the electricity sector by 2030, we will need solutions that keep America economically prosperous and competitive.

As is evident here in the desert of the American Southwest, the sun provides an abundant source of energy. Citizens have been using this source for years to power many small and large-scale projects and devices. Indeed, an entire small community of astronauts live on an International Space Station that is run entirely on solar energy. Yet, solar energy makes up a very small proportion of our overall consumption picture. Only 0.4 percent of global energy demand is met by geothermal, solar, and wind energy *combined.* In the U.S., statistics by the

Department of Energy from 2004 indicate that solar energy accounted for one percent of the total U.S. energy consumption, and 0.2 percent of our electricity generation.

Part of the reason this resource is not widely used is that sunlight is not constant and focused. The amount of energy generated depends on the time of day, location, and weather conditions. In order to use solar energy for stable grid operations, we need better storage techniques. To this end, I introduced a bill last year, the *Energy for America Act,* which included a provision promoting the research, development, and demonstration of energy storage technologies for electricity transmission and distribution. I am pleased that this provision was included in the Energy Bill signed into law last year.

I look forward to hearing the testimony today by this esteemed panel of individuals on how America can better harness the power of the sun to generate more utility-scale power that lowers the cost to taxpayers and consumers.

I yield back the balance of my time.

Ms. Giffords. Thank you, Mr. Hall.

I would now like to yield to the Chairman of the Science and Technology Committee, Mr. Gordon, who will offer his opening remarks (by phone).

Chairman Gordon. Thank you.

I am disappointed as to the mechanical problem that I cannot join all of you in southern Arizona today, but I want to thank you, Representative Giffords, for taking the lead and putting together this important hearing, and I want to thank my colleague, Old Geezer, and the other colleagues there, for being on the scene, and I know you are going to be bringing back a good report for us.

It is obvious that solar energy has the potential to provide a significant amount of power in Arizona, and I look forward to learning more about how states that do not get quite as much sun, like my own State of Tennessee, might be able to benefit from the tremendous resource we have in the southwest.

And again, I would point out that in 2006 Germany installed about seven times more solar power than the entire U.S., and that Germany's solar resources are roughly equal to Alaska's. So, I know that we can be doing much more to utilize the sun's energy.

It is clear that a major component of any scheme to use solar power on a large scale has to be energy storage. So, I am encouraged by the grand solar plan that Mr. Hansen will be describing. Additionally, further development in advanced batteries will also be a critical part of the distributed generation system that Ms. Rauluk will talk more about.

I am pleased about the bipartisan work last year on the energy storage, which had official contributions in this area from both you, Madam Chair, and Mr. Hall, as it was introduced in the latest energy bill and became law in December.

An improved transmission system is also needed, especially, if we ever expect to get a large fraction of our energy from remote regions where renewable resources, like solar and wind, are concentrated.

I am also concerned about the nexus between water and energy, something that we are going to be looking into more this year on the Committee. While regular solar panels do not need much water, except to clean them on occasion, some estimate that solar thermal technology uses more water than a typical coal plant. So, it is important that while we move forward we take the whole picture into account, and do everything we can to avoid trading one big problem for another.

And, I am excited about the opportunities that large-scale solar power present to create thousands of new green jobs, and reduce our dependency on old sources of energy. Our committee will continue to do everything we can to help overcome the barriers to getting us there.

Representative Giffords, thank you again for your strong leadership to promote solar energy, and thanks to this distinguished panel of witnesses for being here today.

I yield back my time.

[The prepared statement of Chairman Gordon follows:]

PREPARED STATEMENT OF CHAIRMAN BART GORDON

Thank you, Congresswoman, for taking the lead in putting together this important hearing. It is obvious that solar energy has the potential to provide a significant amount of power right here in Arizona, and I look forward to learning more about how states that don't get quite as much sun, like my home State of Tennessee might be able to benefit from the tremendous resource we have in the Southwest.

Then again, I could point out that in 2006, Germany installed about seven times more solar power than the entire U.S., and that country's solar resources are roughly equal to Alaska's, so I know that we could be doing much more to utilize the sun's energy.

It's clear that a major component of any scheme to use solar power on a large scale has to be energy storage. I am encouraged that the Grand Solar Plan that Mr. Hansen will describe would make use of compressed air and thermal energy storage technologies.

Further developments in advanced batteries will also be a critical part of the distributed generation systems that Ms. Rauluk will talk more about in just a few minutes. I'm proud that our committee's bipartisan work last year on energy storage, which had essential contributions in this area from both you and Mr. Hall, was included in the latest energy bill that became law in December.

An improved transmission system is also needed especially if we ever expect to get a large fraction of our energy from the remote regions where renewable resources like solar and wind are concentrated.

Our current system of power lines isn't robust enough to carry large amounts of power from these centers to consumers everywhere. Too much energy would be lost over the long distances between generation and delivery of power.

Studies by Oak Ridge National Laboratory show that new high-voltage direct current lines lose far less energy than existing transmission lines over the same distances. They may also be more reliable and cheaper to build. I look forward to hearing more about the prospects for making these kinds of changes to our electric grid system from this panel.

And I am also concerned about the nexus between water and energy. While regular solar panels really don't need much water except to clean them on occasion, some estimates show solar thermal technology using more water than a typical coal plant.

It is important that, moving forward we take the whole picture into account and do everything we can to avoid trading one big problem for another.

I am excited about the opportunities that large-scale solar power presents to create thousands of new green jobs and reduce our dependence on foreign sources of energy. Our committee will continue to do everything we can to help overcome the barriers to getting us there.

Representative Giffords, thank you again for your strong leadership to promote solar energy, and thanks to this distinguished panel of witnesses for being here today.

Ms. Giffords. Thank you, Mr. Chairman.

Next, I would like to yield to Representative Harry Mitchell, for his opening statement.

Mr. Mitchell. Thank you, Madam Chair.

I would also like to thank Chairman Gordon and Congresswoman Giffords for organizing this field hearing.

As a fellow freshman Member of Congress, from the sunny State of Arizona, Ms. Giffords and I have a unique perspective on how to address our nation's energy crisis.

We are lucky here in Arizona to enjoy over 300 days a year of sunshine. We have a real opportunity to brighten our state's future by investing in solar energy research and technology.

As solar technology advances, I believe that Arizona will be a leader in clean alternative energy production. Refocusing our energy production on alternative sources such as solar is critical for our national security and the environment.

Moreover, investing in solar energy is vital to Arizona's economy. Recently, Arbengoa Solar, and Arizona Public Service, announced exciting plans to develop the Solana Generating Station, and that is a 280 megawatt solar thermal energy plant, right here in the southwest. This will be the world's largest solar power plant. Solana will not only be a leading source of emission-free electricity, but it will also start significant development for Arizona.

However, the Arbengoa and Arizona Public Service executives have candidly told us that the Solana project will not happen without the extension of essential solar tax credits.

I am proud of the work that we have done in Congress to make sure that utility-scale solar projects, in particular, like Solana, continue to benefit from solar tax credits.

Recently, I voted for, and the House passed, the *Renewable Energy Conservation Act,* which would extend the 30 percent investment tax credit for solar energy property for eight years, through 2016. For the first time, public utilities would also be able to claim this investment tax credit.

I remain committed in doing what I can do in Congress to encourage further development and production of solar energy, and I would also like to thank two people from my district who are here to testify today. Barbara Lockwood is the Manager of Renewable Energy for Arizona Public Service, and Kate Maracas is the Vice President of Arbengoa Solar's Arizona Project. Both Arizona Public Service and Arbengoa are leaders in utility-scale solar energy and are working together to develop the Solana plant.

I look forward to hearing more about what we can do to establish Arizona's reputation as the Solar State, and I yield back the balance of my time.

Ms. Giffords. Thank you, Mr. Mitchell.

I would like to yield a few minutes to Representative Matheson for his opening statement.

Mr. Matheson. Well, thank you, Madam Chair. I will be very brief, because I am looking forward to hearing from the panel. But, I just do want to emphasize that the Science Committee is a very bipartisan committee in the House of Representatives. It is a committee that usually takes on a longer-term view on issues, and I think the issues before the Science Committee are really the issues that matter when you look out a couple of generations from now.

Congresswoman Giffords, you have been a leader in advocating the solar energy issue. You are a great Member of the Science Committee, and I am really impressed with how you have brought this hearing together today, and I just want to acknowledge that in your freshman term you have already established yourself as a real champion for this issue, and I will yield back my time.

Ms. Giffords. Thank you, Mr. Matheson.

Representative Lipinski.

Mr. Lipinski. I just want to very briefly say, back in 1980 I was in 8th grade. We had a science fair. I did a science fair project on solar energy. It seems like in the—well, soon afterwards, we stopped having much interest in solar energy. It seemed like 30 years ago that was the emerging energy technology, where we had all these concerns about high gas prices, what is going to happen with oil, our oil supply, but here we are 30 years later, it feels like we are in the same place.

It is very important that we do not make mistakes now that we made back then. So, that is why I am very happy that Congresswoman Giffords is holding this hearing today. You know, having this many Members of Congress come out to a field hearing just really shows how important the issue is, and the role that Congresswoman Giffords is playing here—you know, this is a freshman Member of Congress—to have all this out here.

As Vice Chairman of the Science Committee, I think this is just a fantastic opportunity that we have here today. This is critical for the future of our country, and it is very important, critical for southern Arizona, certainly, and I look forward today to hearing from all the witnesses here today, and making sure that we do everything we can at the federal level so that we can take advantage of what is available, what solar energy makes possible for us.

And, I know Congresswoman Giffords is going to be a leader in Congress in doing that.

So, I yield back.

Ms. Giffords. Thank you, Mr. Vice Chairman.

[The prepared statement of Mr. Smith follows:]

PREPARED STATEMENT OF REPRESENTATIVE ADRIAN SMITH

Thank you, Madame Vice Chairwoman. Nebraska is a state blessed with many natural resources, not least of which is sunshine. I have long held, and continue to believe, the United States needs to explore a diverse array of energy technologies.

Solar energy is one of many technologies which hold exciting potential. It is one of our most ubiquitous and reliable resources. Although there are challenges associated with capturing and storing solar energy, they are not insurmountable. I believe solar energy will play an important role in our future energy security.

I support policies which encourage innovation, research, development, and investment in renewable energy. We need to encourage long-term investments in solar energy, as well as wind energy, biofuels, nuclear power, and hydro-power. Our own domestic oil and gas resources should not be overlooked.

Thank you, Madame Vice Chairman, and I look forward to working with you to further policies which will promote research, development, and investment in solar technologies and other energy resources, leading to better energy security and national security for every Nebraskan and every American.

Ms. Giffords. Now we would like to hear from our panelists, our witnesses. We are going to start, with just a couple brief seconds of introduction. We are going to hear from Mr. Mark Mehos, who is the Program Manager for the Concentrating Solar Power Program at the National Renewable Energy Lab in Colorado. So, thank you so much for being here.

We are going to hear from Mr. Tom Hansen, who is Vice President of Environmental Services, Conservation and Renewable Energy, at Tucson Electric Power, also featured in the *''ScientificAmerican Magazine''* article, so great to have you here as well.

Ms. Kate Maracas is the Vice President of Arizona Operations at Abengoa Solar. We have heard a lot about your company and the proposed project. We are excited that you are here today as well.

Ms. Valerie Rauluk is the Founder and CEO of Venture Catalyst, Inc. She is a passionate supporter of solar energy, but also speaks on behalf of the business community as well. So, Ms. Rauluk, thank you for being here.

Ms. Barbara Lockwood is the Manager of the Renewable Energy for Arizona Public Service, APS, thank you for coming down from Phoenix, we welcome you.

And finally, Mr. Joe Kastner, the Vice President of Implementation Operations for MMA Renewable Ventures, LLC. We are so pleased to have you in southern Arizona. Thank you so much.

As I said earlier, our witnesses will have five minutes to present your oral testimony. Your written testimony will be inserted into the record, and when we are finished with the witnesses' testimony, remember that we will take turns asking questions of our witnesses.

And, we will begin with you, Mr. Mehos, please.

STATEMENT OF MR. MARK MEHOS, PROGRAM MANAGER, CONCENTRATING SOLAR POWER PROGRAM, NATIONAL RENEWABLE ENERGY LAB, COLORADO

Mr. Mehos. Okay, thank you, Madam Chairman, thank you Members of the Committee, for giving me the opportunity to speak today. I did provide written testimony and ask permission to provide a Power Point oral presentation. Much of my information is better viewed than discussed.

So, I was asked to present information on, basically, some background on utility-scale solar power. I will do that. At NREL we have done a lot of analysis on the overall resource side for utilityscale solar generation in the U.S., especially with an emphasis on the southwest.

I was asked to present information on what the Federal Government can do to facilitate deployment, and finally, what can the Federal Government do to kick start utility-scale solar.

Real quickly on some background of utility-scale solar technologies. When we talk about utility-scale solar we are really talking about two different markets, solar with storage, which we call dispatchable generation, which I will discuss in a little bit, and solar without storage or, basically, non-dispatchable, meaning that it will generate electricity when the sun is shining.

On the dispatchable storage, we are really talking about three categories of technologies, and there are variations within these that I do not have time to discuss, but there's the parabolic trough up on the top of the screen, the power tower over to the lower left, and finally the linear fresnel, which is an upcoming technology over on the lower right. Each of these technologies are thermalbased technologies, which means they use concentrated sunlight to generate heat that can then be converted to electricity using conventional steam sites.

The solar without storage, or what we call the non-dispatchable technologies, are just that. They are, basically, they cannot use storage, and there is three technologies that can fall into this category, the dish engine, which is on your upper, or on the lower side of the screen now on your left, concentrating photovoltaic technologies, which is in the middle, and then finally flat plate PV technologies, just like those that are on the home, but in a much larger scale for utility-scale applications.

On the top, most of these are built in very large installations, on the order of 50, 100, 200 megawatts. People are even talking about much larger installations. Their economies of scale forced them toward that side. On the lower part of the screen, the nondispatchables, typically, those are much smaller-scale technologies, on the order of kilowatts to tens of kilowatts, which can then be gained together to the similar 100 megawatt plants for large utility-scale applications.

So, I talked about the dispatchable technologies. Why do we, or why do utilities, like dispatchable power? Primarily, what we do with dispatchable power, is we collect a lot of that energy that is shining on our collectors during the daytime, we store a lot of that energy, or most of that energy, and we distribute that energy over a larger period of the day. So, what does that do for us? It gives us a much higher value for that energy collected. If we were just dispatching during the peak times of the day, that is a very high value period, certainly, but especially in the southwest when people come home and they turn on their air conditioners, and their TVs, and everything else, that peak load really does go throughout later into the evening and even into the night. And so, we are allowed to take on that higher value production later into the day, not just during the daytime. It also allows us to lower the cost of the technology, basically, taking what amounts to be a fairly high capital cost for any solar technology and amortizing that high capital cost over a larger number of hours, instead of operating annually at 25 percent of the year, we can operate up to 50 percent or even 70 or 80 percent of the year. So, it allows you to really take that energy and lower the cost, due to that larger amortization.

Okay. So, you asked the question regarding what is the resource potential, so at NREL we have done a significant amount of analysis using geographic information systems as a start, to try to screen where the most economical locations for large-scale solar can make sense.

Up in the upper left-hand corner, I show the unfiltered solar resource, basically, the darker red spots are where the highest solar resource exists. But, if I take this entire southwest region and compare that southwest region to anywhere else in the world the southwest is as high or higher than anyplace else in the world as far as its solar resource.

But, that is not all we need to look at. We need to look at other exclusions that can lower the economic value of that. The first thing we want to do, on the upper right-hand corner, is to exclude those areas that have a lower solar resource. We picked six hours, six kilowatt hours per meter squared per day, not that in the long-term you could not build economic plants, but we were looking more in the near-term.

On the lower left-hand corner, we start looking at land exclusions. This is utility-scale solar. Typically, plants can be as large as one square mile, for the APS plant three square miles, so we are looking at large tracts of land that we are not going to build in wilderness areas, we are not going to build in urban areas, and so we take those types of exclusions, and that is the lower left-hand corner.

Finally, because we are trying to minimize costs for constructing the plant, we are looking at fairly flat land. That does not have to be the case for some technologies, the most aggressive case is in the lower right-hand corner where we have looked at just one percent slope land.

So, when we do that, and we look at that fairly aggressive filtering scenario, we still have 11 times the current U.S. generating capacity, just in that small percentage of land that is left over. The current U.S. capacity is about 1,000 gigawatts, and I am showing a solar capacity with those lands of 11,000 gigawatts of potential.

Okay. So, you asked what can the Federal Government do to support this technology. I should say on that last map there are a couple items in my written testimony. One is, a lot of that land that I showed, while a lot of that lies on private lands, much of that lies on federal lands, on Bureau of Land Management lands, on Defense lands, so there is an effort within the Department of Energy to try to open access, to try to reduce barriers to permitting, for example, to putting some of these large scale solar plants on federal lands. And so, the Federal Government can continue to support that effort.

Also based on that map, much of that solar resource, while it is located near existing load, located near existing transmission, much of that transmission is constrained, and so to the degree that the Federal Government can support regional efforts to try to extend that transmission into these high-value areas, whether it is solar, or whether it is wind, or other renewable resources, please continue that effort.

Probably one of the most important near-term issues facing large-scale solar is the extension of the 30 percent investment tax credit. So, this is a quick example of the cost reduction associated with going from a 10 percent permanent investment tax credit to the 30 percent investment tax credit, basically, about a 15 percent reduction in cost, which does not seem like an awful lot, but this technology is right on the margin, as I will show in a second, that 15 percent reduction in cost to make or break these systems.

So, this is a curve based on a regional deployment system model developed at NREL, that competes solar technologies, wind technologies, against conventional technologies. I will not go into detail, but without an extension of the existing 30 percent investment tax credit, in other words, just the 10 percent investment tax credit, we see very little new penetration of utility-scale solar technology in the near-term. You do see it come on line in the long-term, as conventional costs start to rise, and as our costs start to reduce based on the R&D that happens within the laboratory system.

If you do extend that 30 percent tax credit by eight years, and it is important to note that a two-year extension does nothing, and eight-year extension will allow, according to our models, and we hear this out in the public sector, quite an increase, up to 20 gigawatts in the near-term of this technology. So, the extension of the tax credit is extremely important.

This just gives you a more visual representation of that, without the extension of the tax credit, and our model does go to 2050, but this is the nearer-term look. You will get some initial penetration that is basically driven by some of the State portfolio standards, and that initial penetration will probably happen 10 or 15 years from now, but you will get some initial penetration, and this, basically, the colors show capacity in megawatts. So, low penetration.

If you were to extend this investment tax credit by eight years to 2020, in this case I am showing the data, then you are looking at significant penetration in Arizona, in California, down in Texas, New Mexico, Colorado, all of these states driven, primarily, by the extension of this tax credit.

You asked what would be needed to kick start utility-scale solar. Right now, within the DOE program, utility-scale solar is about $30 million, primarily, focused on concentrating solar power of $170 million budget, which is mostly photovoltaic and distributed generation.

NREL and Sandia, at the request of DOE, did an exercise to see what could happen to accelerate utility-scale solar, and we estimated about $50 million per year to achieve accelerated goals. Those accelerated goals included being competitive in intermediate load markets by 2015, a five-year acceleration of the current goal, and to be competitive in carbon-constrained base load markets by 2020, which was not a goal previously at all.

To achieve those goals, there needs to be R&D emphasis on advanced thermal storage systems, and higher temperature systems, primarily, the troughs and the tower systems, which are the systems that are high temperature and allow for thermal storage.

I believe that's the end of my presentation.

[The prepared statement of Mr. Mehos follows:]

Prepared Statement of Mark Mehos

Madame Chairman, thank you for this occasion to present and discuss information related to opportunities and obstacles for utility-scale solar power. I am the Manager of the Concentrating Solar Power Program at the National Renewable Energy Laboratory (NREL). NREL is located in Golden, Colorado, and is the U.S. Department of Energy's primary laboratory for research and development (R&D) of renewable energy and energy efficiency technologies. I am honored to be here and to speak with you today.

I truly believe that solar power—both concentrating solar power and photovoltaic technologies—can provide a significant level of generating capacity in the United States if cost goals established by the U.S. Department of Energy (DOE) can be achieved. Reaching these goals will require a carefully balanced blend of DOE and industry sponsored R&D and government policies.

Introduction to Solar Technologies

Solar energy can be converted into electricity by means of photovoltaic (PV) or concentrating solar power (CSP) systems. Photovoltaics is the technical word for solar panels that create electricity. Photovoltaic material converts sunlight directly into electricity through a device called a solar cell. When sunlight strikes a solar cell, electrons are dislodged, creating an electrical current that can be captured and harnessed to do useful work.

Solar cells are connected together electrically to produce modules, and modules are mounted in PV arrays that can measure up to several meters on a side. **Flatplate PV** arrays can be mounted at a fixed angle facing south, or they can be mounted on a tracking system, allowing the array to follow the sun in one or two axes to capture more sunlight over the course of a day. About 10 to 20 PV arrays can provide enough power for a household. However, for large electric utility or industrial applications, hundreds or thousands of arrays can be interconnected to form a single, large "utility-scale" PV system.

Higher efficiency solar cells, because of their high cost, are better suited to operate under concentrated sunlight. **Concentrating photovoltaic** (CPV) collectors use lenses or mirrors as optics to focus the sunlight onto the high-efficiency cells. The main idea is to use very little of the expensive semi-conducting PV material while collecting as much sunlight as possible with lower-cost concentrating optics. CPV systems are being considered primarily for utility-scale applications.

CSP technologies use concentrating optics to generate high temperatures that are typically used to drive conventional steam or gas turbines. Due to economies of scale, CSP is generally considered a central-generation technology, rather than a source of distributed generation.

The three main types of concentrating solar power systems are parabolic trough systems, power tower systems, and dish/engine systems. Variants of these systems are also being considered, such as the linear Fresnel reflector system, which uses flat, rather than parabolic, mirrors to concentrate the solar thermal energy.

Parabolic trough systems concentrate the sun's energy through the use of long, linear parabolically curved mirrors. The mirrors track the sun, focusing sunlight on a receiver that runs along the focal line of the trough. A heat-transfer fluid, typically a synthetic oil, flows through the receiver, rising in temperature as it flows along the length of the collector. The hot oil is then used to boil water in a conventional steam generator to produce electricity. Alternatively, water can be boiled directly in the receiver using a direct-steam receiver. A key advantage of parabolic trough systems is that they can use thermal storage, giving the systems the flexibility to dispatch electricity coincident with peak utility loads, which often occur late in the evening. Many systems in Spain, as well as the system announced by Arizona Public Service last month, will make use of this feature. Parabolic trough systems are currently the most commercially developed technology.

A **power tower** system uses a large field of mirrors, called heliostats, to concentrate sunlight onto the top of a tower, where a receiver is located. This focused sunlight heats a working fluid such as molten salt or water/steam flowing through the receiver. Similar to oil in a parabolic trough receiver, the salt in a tower receiver is used to generate steam (using heat exchangers) to generate electricity through a conventional steam generator. As with trough systems, tower systems can be integrated with thermal storage. Future low-cost storage options should allow both troughs and towers to operate competitively in the near-term in intermediate load markets and in the future in base load markets, offering a potential alternative to coal-based generation.

A **dish/engine** system uses a mirrored dish, similar to a very large satellite dish. The dish-shaped surface collects and concentrates the sun's heat onto a receiver, which absorbs the heat and transfers it to a gas within a Stirling engine or gas turbine. The heat allows the gas to expand against a piston (in a Stirling engine) or to power a turbine to produce mechanical power. The mechanical power is then used to run a generator or alternator to produce electricity.

Resource Potential for Solar Energy in the United States

A 2005 study commissioned by the Western Governors' Association (WGA) looked at the solar resource and suitable land available in seven southwestern U.S. states, including California, Arizona, Nevada, Utah, Colorado, New Mexico, and Texas. Analysis using Geographic Information Systems (GIS) determined optimal CSP sites with high economic potential by excluding regions in urban or sensitive areas, regions with low solar resource, and regions where terrain would inhibit the cost-effective deployment of large-scale plants. Even with this high level of exclusions, the WGA solar task force calculated a capability of generating up to 6,800 gigawatts (GW) using CSP technologies—almost seven times the current electric generating capacity of the entire United States. The WGA study found that, with a build-out of only two to four GW of CSP, the technology will be competitive with conventional natural-gas-fired combined-cycle plants with a cost approaching 10 cents per kilowatt-hour.

The southwestern United States is not the only area with great potential for CSP. Projects are under way in Spain and Northern Africa, with additional projects planned for Israel, the Middle East, Northern Mexico, and Australia. In total, more than 60 utility-scale CSP plants are under development worldwide, primarily driven by policies favorable to large-scale deployment of the technology.

Two questions are now addressed that relate to the role of the Federal Government in the success of utility-scale solar projects in the Untied States.

The first question is: How can the Federal Government Facilitate the Deployment of Utility-scale Solar Projects?

At the request of the U.S. Department of Energy, NREL analyzed the impact of policy (both State and federal) and R&D on the penetration of utility-scale solar generating systems in the southwest United States. The Renewable Energy Deployment System (ReEDS) model, developed at NREL, was used to estimate the U.S. market potential of wind and solar energy for the next 20 to 50 years. The model competes these technologies against the more-conventional generation technologies of hydro, gas-combustion turbine and combined-cycle systems, coal, and nuclear. Future sequestration technologies are also included within the ReEDS model.

Results from the model indicate that utility-scale solar technologies can produce nearly 120 GWs of capacity in the Southwest by 2050. Significantly more capacity is possible if dedicated transmission can supply generation to load centers located outside the Southwest. However, a key outcome of the analysis is that initial market penetration is extremely dependent on the continuation of the existing 30 percent investment tax credit (ITC). According to the analysis, without an extension of the ITC, new capacity will be delayed about 10 to 15 years—until lower CSP generation costs resulting from R&D and international market development allow CSP technologies to compete against future conventional plants.

The Federal Government can facilitate the deployment of solar power plants by providing access to land. Utility-scale solar projects require considerable acreage. The 280-megawatt (MW) Arizona Public Service project mentioned earlier will cover three square miles. That is nearly 2,000 acres to produce the power for 70,000 homes. The Federal Government owns large tracts of land in the West. Doing an environmental study of those lands and streamlining the process by which industry can lease tracts found suitable for solar power projects will shorten the time it will take to build projects on these lands.

Finally, the Federal Government can support efforts to relieve transmission congestion throughout the West. Existing transmission lines are operating at near capacity. New lines must be built to bring power from solar plants located in the areas where the solar resource is best, often in remote sunny regions. Our transmission grid is like our highway system, but without the interstate highways. An ''interstate'' grid system would facilitate the transmission of solar power from the Southwest to load centers throughout the United States. As described earlier, the United States has an enormous solar resource. Once we reduce the cost of the technology, the next challenge will be to distribute the electricity produced to the people who need it.

A second Questions is: How does the Level of Federal Investment Required to ''Kick Start'' Utility-scale Solar Compare with that Required by Other Technologies Seeking Government Support?

NREL scientists are studying a number of renewable energy technologies. The country is entering a period where it must start making the transition to new sources of energy. DOE and NREL are pursuing a portfolio approach of technologies —such as solar, wind, biomass, hydrogen, and geothermal—that could play a role in the future. All these technologies have the potential to become cost competitive with fossil generation.

The price tag of utility-scale solar projects is large. For example, the 280-MW plant mentioned above will cost more than a billion dollars. Fortunately, the Federal Government does not need to contribute directly to cover the cost of these plants. The southwestern states have established renewable portfolio standards that have created the market for utility-scale power plants. Some states have established price guidelines by which they recognize that they will initially have to pay more for the renewable power. The additional costs are passed along to the rate payers. Thus, the bulk of the cost for establishing cost-effective utility-scale solar power is being borne by the states. If the Federal Government were to decide that utility-scale solar power was important, then they could partner with the states, which have already kick-started utility-scale solar.

Most of the money appropriated for solar energy R&D focuses on residential and commercial applications. Utility-scale solar receives about $30 million out of a total solar budget of $170 million. To meet the goals mentioned earlier, the DOE estimates that this funding would need to be doubled. Researchers at NREL work closely with the CSP industry and universities to develop new technologies that are more efficient and less costly. A study commissioned by DOE several years ago showed that reducing the cost of solar technology depends about 45 percent on R&D and 55 percent on actually building solar projects. This combination of R&D and deployment could well bring the cost of solar power into alignment with fossil generation in the intermediate power markets. And with low-cost storage, the overall cost may also align with future baseload power markets if carbon constraints are considered.

Summary

Addressing our near-term needs in solar power will require a national strategy that promotes the deployment of solar systems and processes that are ready to serve us today. At the same time, addressing our longer-term needs and achieving a significant contribution from solar power technologies will require a major new commitment to the research needed to deliver the next—and subsequent—generations of CSP, PV. and other new solar technologies.

Thank you.

Ms. Giffords. Thank you, Mr. Mehos.

Mr. Hansen, please.

STATEMENT OF MR. THOMAS N. HANSEN, VICE PRESIDENT, ENVIRONMENTAL SERVICES, CONSERVATION AND RENEWABLE ENERGY, TUCSON ELECTRIC POWER

Mr. Hansen. Good afternoon, Chairwoman Giffords, thank you for the opportunity, thank you, distinguished Members of the Committee, for the opportunity to be here today to discuss with you what I consider a very important solution and a plausible solution to our energy challenges for the future, the ''Solar Grand Plan,'' as described in the ''Scientific American'' issue in January of 2008.

While many energy resources are sustainable and environmentally neutral, only solar energy can supply all of our projected energy needs for the long-term sufficient needs to meet our longterm needs in the United States for centuries to come.

The January, 2008 issue of *''Scientific American Magazine''* details that ''Solar Grand Plan,'' and it could realistically provide 69 percent of U.S. electric needs by 2050, and 100 percent of U.S. electric needs by the year 2100.

I should point out that my comments do not necessarily represent the views of Tucson Electric Power (TEP), although I'm employed by TEP these comments are based on my thoughts and my development of a totally alternative energy system over the last decade. So, TEP has not, at this point, taken a position on this ''Solar Grand Plan.''

The three core components of the ''Solar Grand Plan'' are first, solar energy installations, of course, to convert the sunlight into electricity. The second component, and a very important component, is energy storage. We need to be able to store that energy from the sunlight to make it dispatchable for periods when the sun is not shining. And finally, electric transmission, in order to bring the power from the southwest, or from other sources, be they hydro sources, or hydro kinetic, ocean sources, wind sources, throughout the United States, to the customers and to the energy storage sources.

The ''Solar Grand Plan'' assumes the deployment of a combination of thin film photovoltaic technologies, about 88 percent, and thermal storage, thermal solar technologies, including thermal storage, about 12 percent. These technologies have a proven record of reliability and safety, and that is what I want to emphasize, the ''Solar Grand Plan'' uses proven technologies that are in existence today. They are a little more expensive than traditional technologies, but they are proven. They do work.

The production of solar energy is a function of the time of day, time of year, and cloud cover. We need to have storage to be able to mitigate those intermittencies to provide utilities with a tool to make consistent dispatchable electricity. Solar energy has been developed, the modules themselves, the solar collectors of the thermal solar systems have been developed to a very high degree of reliability. What we need now is an effort to provide that storage, and to link the storage elements together with transmission opportunities.

One great reason that we would propose using compressed air energy storage is because it is under ground. It uses the same technologies as have been used to develop natural gas storage, and, in fact, only uses about 10 percent of the capacity that has been used already to develop natural gas storage in the United States. Energy storage under the ground makes it more secure, makes the opportunity to have it widely disbursed, providing better opportunities for energy security in the United States in the future.

Finally, the transmission component. In order to tie everything together, to make the opportunity to move electricity, just as we have in the United States with our interstate highway system, we need to have an interstate transmission system. This will probably be, while it is, again, one of the more technically, easily implemented elements, it will be probably the most challenging element from the standpoint of regulations and the ability and need to work together with stakeholders from states, at the federal level, and local level, to be able to implement putting in these right-of-ways for transmission systems.

There are other technologies as well that need to be developed. The smart grid technologies, as we heard, are a very important part. They make the glue, if you will, to tie all the pieces together. But, much, again, of that work has been done already. These are nascent technologies, but they do exist. They are not far fetched, this is not non-existent technology.

With all energy project proposals there is a price. This program should take, we estimate, about $420 billion, but think of it as another opportunity, another alternative, to carbon collection and storage. As we start looking at alternatives towards our energy future, we need to weigh the different alternatives. Carbon collection and storage is one opportunity, but so we feel is also the ''Solar Grand Plan.''

And finally, the ''Solar Grand Plan'' strengthens our energy security by effectively storing electrical energy under ground, just as we have done with fuels at the Strategic Petroleum Reserve and at natural gas storage facilities.

And, just as it took the political will of Congress, with strong support from the states, to make the interstate highway system a reality, so it will take strong leadership, vision and the will of Congress, in partnership with the states, to make this become a longterm, reliable, sustainable, energy secure solution for our nation.

Madam Chair, I thank you for the opportunity to visit with you today, and I look forward in the future to questions.

Thank you.

[The prepared statement of Mr. Hansen follows:]

Prepared Statement of Thomas N. Hansen

Thank you very much Chairman Lampson, Vice Chair Giffords, Ranking Member Inglis and distinguished Members of the Committee and staff. My name is Tom Hansen and I am the Vice President of Environmental Services, Conservation and Renewable Energy for Tucson Electric Power, the second largest investor-owned electric utility in Arizona. We serve the energy needs of nearly 400,000 customers in the Tucson area. It is always a great honor and pleasure to work with the Members of Congress and their staff in exploring solutions to the energy challenges facing Americans today. The production of affordable, safe electricity from sustainable and secure sources in an environmentally appropriate manner is one of those challenges. I am here today to discuss with you one plausible solution to that challenge —a Solar Grand Plan.

My background includes the design, construction, operation or management of over 10,000 MW of generation capacity, comprised primarily of nuclear and coal power plants. Many of my solar advocate friends claim I am serving my penance for that background by working with solar energy. But I am proud to have played a role in developing the electric generation infrastructure that has served our society so well for many decades and will play a significant role in providing needed electricity for at least the next twenty years. Moreover, my background in the development of those traditional electric generation assets has given me a unique insight into the technical and operational characteristics of the next generation of power producing technologies. While many energy resources are sustainable and environmentally neutral, only solar energy can supply all of our projected energy needs for the long-term future. Solar energy is abundant, ubiquitous, sustainable and sufficient to meet the total energy needs of the United States for centuries in the future. The January 2008 issue of *Scientific American* detailed a ''Solar Grand Plan'' that could realistically meet 69 percent of U.S. electric needs by 2050 while reducing electricity related greenhouse gas releases by nearly 50 percent. The same plan could satisfy 100 percent of our nation's electric needs by 2100. I am here today to discuss that plan with the Committee.

My enthusiasm for the Solar Grand Plan will likely become obvious throughout the course of this testimony. Nevertheless, I should point out that my comments do not necessarily represent the views of Tucson Electric Power. TEP has earned widespread recognition for its innovative solar power programs, and the company is committed to expanding its use of renewable power resources. But TEP has not taken a position on this particular plan for solar power development.

The Solar Grand Plan incorporates three core technologies that will be coordinated through smart grid technologies. The smart grid, which will be discussed in more detail later, would be a bi-directional quasi-real time communications and control system with interconnected energy sources, including traditional, wind, tidal, hydrokinetic, biomass and geothermal. It would connect with consumer electric devices —including plug-in hybrid electric vehicles—and incorporate predictive solar and wind forecasts of both short-and long-term time spans. While this system is incorporated in the Solar Grand Plan, the technology has enough flexibility to seamlessly integrate any traditional fueled or sustainable energy resource.

The three core components of the Solar Grand plan are:

- Solar generation to refine the energy in the light rays of the sun into electricity.
- Energy storage to preserve energy from the sun or any other power source for later conversion into electricity.
- Electric transmission to link the solar generation and energy storage to consumers and their equipment.

Discussion of each core component follows:

Solar Generation

The solar generation component converts the energy of the sun's rays into the electric energy that will be stored or delivered to customers. Any of a wide variety of existing solar energy technologies will meet the requirements for this component of the Solar Grand Plan, as the final output of all solar electric technologies is effectively interchangeable. Fixed or tracking flat plate, crystalline or thin film photovoltaic (PV) modules with DC-to-AC inverters for interconnection to the electric grid are supported. Concentrating solar power (CSP) using trough, solar tower, dish/thermal engine, dish/PV engine or concentrated photovoltaic (CPV) technologies are all supported.

Tucson Electric Power's solar development experience has demonstrated that deployment of a varied portfolio of solar technologies is needed for optimal solar generation economics, as each solar technology type is best suited for operation in a particular set of climatic conditions. For example: PV, which uses virtually no water in operation, is best suited to a geographic area with no access to water; solar trough technology, meanwhile, is appropriate for areas with readily available water and can be combined with desalination options to augment the local potable water supply. The Solar Grand Plan assumes the deployment of a combination of thin-film PV technologies (88 percent of the total) and solar thermal with storage technologies (12 percent of the total). Both technologies are commercially available and are in common use today, with a very good opportunity for future cost reductions.

The solar generation technologies envisioned in the Solar Grand Plan have a proven record of reliability and safety. Future challenges for these technologies include reducing their cost through development of larger-scale U.S. located manufacturing facilities, optimizing the balance of system component costs and standardizing installation code requirements. Solutions to these challenges include an extension of the federal Investment Tax Credit for solar energy systems, expanded support for research and development of new solar energy conversion technologies at federal laboratories and universities, continued support for the Solar America Initiative program goals, and expanded federal programs for education and outreach to the American people regarding developments in solar energy product commercialization.

Energy Storage

A core concept of the Solar Grand Plan is the need to store solar energy for use when the sun is not shining. By way of explanation, traditional electric generation is performed by technology that effectively refines a form of primary energy such as chemical energy in coal or nuclear energy in uranium into electricity. Instantaneous customer demand for electricity is met through the conversion of just the right amount of primary energy source to perfectly balance the supply of electrons with demand for those electrons at all times. A utility's ability to meet its customers' peak electric demand is a function of the maximum capacity built into the power plant.

Production of solar energy—effectively refining the energy of the sun into electricity —is a function of time of day, time of year and cloud cover as well as capacity of the plant. The energy of the sun at the Earth's surface is not dispatchable, to use a traditional electric system term, as the utility has no direct control over increasing the sun's intensity. The addition of cost- effective, reliable, efficient, safe, environmentally compatible energy storage into a solar generation system would allow a utility to control output to support customer loads during times when the sun's energy is not available. Energy storage is needed on at least two different time scales: storing excess solar energy in one season for consumption a season or two later, and storage of daytime solar output for use at night.

Numerous energy storage technologies exist today, but only two are suitable for utility scale use. One of them, pumped hydro storage, retains energy in the form of potential energy in water stored at a high elevation. The energy is released when the water is allowed to flow to a lower elevation through a traditional hydro generator. The other option, compressed air energy storage (CAES), retains potential energy in the form of high-pressure gas in an underground cavern or pore structure. While pumped hydro requires a specific set of geographic surface features and a supply of water, the underground conditions that would allow CAES are available in nearly every state in the union and often are used today for natural gas storage. The same technologies used for developing natural gas storage would support CAES development. The compressors that would tap excess solar generation to pump air underground are available today, as are the combustors and expansion turbines needed to convert the energy in compressed air to electricity. Such generators effectively would employ a split-shaft combustion turbine, thus relying on the same technology used to generate power from fossil fuels today. Two CAES plants are operational today with decades of operating experience, and other plants are in development for use in balancing the output of wind generation.

The Solar Grand Plan requires a total underground storage volume of less than 10 percent of the volume used today for storage of natural gas, a very feasible amount of underground development using existing technology. Surface disturbance area for CAES is very similar to that of natural gas storage, with an additional need for electric transmission access. Energy storage cycle efficiencies of CAES are around 80 percent using existing technologies. While existing CAES technologies use some natural gas to reheat the air prior to conversion of the stored energy back to electricity, adiabatic expansion turbines can be developed that will not require the use of natural gas during energy recovery. A comparison of CAES characteristics with those of other energy storage technologies typically associated with solar energy, such as batteries or flywheels, is very favorable for CAES. Existing battery technologies typically can support a limited number of storage cycle before they must be replaced. Meanwhile, storage of energy from one season to another in batteries or flywheels generally results in additional storage cycle losses dependent upon the duration. CAES can store energy with minimal loss for extended time periods, and capacity does not degrade with an increasing number of charge/discharge cycles.

Meeting all of the Nation's electric needs with solar power would require a sizable volume of energy storage to manage the seasonal and daily variations of solar energy production. While this storage could involve pumped hydro, flywheels, supercapacitors or superconductive magnetic energy storage (SMES), CAES enjoys a significant advantage in that it relies on deep underground facilities. This reduces its impact on land use, limits direct human interaction with the stored energy and mitigates risk from attack by enemies of our country. While the Solar Grand Plan can use any type of efficient, low-cost energy storage system, CAES was chosen due to its existing commercial availability, relatively low cost and proven reliability in utility scale applications. Thermal energy storage in molten salts, a relatively new technology incorporated in plans for thermal solar generation plants, could become part of the Solar Grand Plan's short-term storage component if its reliability and cost effectiveness prove comparable to those of CAES.

Again, the Solar Grand Plan has sufficient flexibility to accommodate any new storage technology that can improve upon the reliability, efficiency, low environmental impact and cost effectiveness of CAES. Development of the energy storage system could be supported by federal Investment Tax Credits to reduce the effective initial cost to the owner of an energy storage facility; supportive capacity tariff rates from regional transmission organizations; and continued support of energy storage technology research and development, including additional funding for evaluation of geologic potential for CAES throughout the United States through the National Laboratories and universities. Favorable regulatory policy ensuring that the requirements of permitting a CAES facility are no more complex than permitting natural gas storage also would help reduce obstacles to development of the energy storage needed for the Solar Grand Plan.

Transmission

The Solar Grand Plan includes a transmission component to collect and distribute the energy produced at the solar generation sites to the energy storage component and to energy-consuming customers throughout the Nation. The Solar Grand Plan does not specifically mention support for wind, biomass, hydro, tidal, current and other forms of sustainable energy production. Nevertheless, it would provide a national transmission backbone—similar in nature, if not in scope, to the National Interstate Highway system—capable of carrying energy from any generation source. The use of high-voltage DC lines for the Solar Grand Plan would leverage proven technologies to significantly reduce the risk of technical problems. However, the development of a national electricity backbone to enable the delivery of energy produced anywhere in the U.S. to any other part of the country will require the same sort of sustained political will that supported the 35-year effort to complete the Interstate Highway System. Policy development will need to address the concerns of property owners, regional and State development priorities and special interests. We must seek to forge a coordinated set of transmission development incentives that will ensure the full acquisition of the right-of-way required for completion of the transmission backbone in a timely manner.

Funding will be needed for research analysis and development of the specific right-of-way alignment for the transmission backbone system. Strategic ''off-ramp'' locations for interconnection to existing regional and local transmission systems will need to be determined through extensive review and analysis of existing available transmission capacity. Cost-effective solutions for transmission bottlenecks to ensure efficient transfer of energy throughout the Nation will need to be found through additional analysis by National Laboratories and universities in cooperation with regional transmission organizations and electric utilities. All of the high-voltage DC transmission technology exists now to make the transmission backbone a reality. Nevertheless, development of a national electric transmission backbone will be the most difficult component of the Solar Grand Plan to implement because of the need to resolve a myriad of permitting and regulatory issues.

Additionally, rules to allocate the system's costs to regions and various customer groups will have to be finalized prior to implementation of the transmission backbone. Technical advancements in high-temperature superconductors could make the regulatory challenges easier to resolve by providing an option for burying the transmission system underground in congested areas where overhead line extensions could be unacceptable. Federal Investment Tax Credits for companies investing in the transmission backbone would offer financial incentives for attracting investment in the transmission backbone system.

Smart Grid

To maximize performance under the Solar Grand Plan, there is a need for a communications and control system to coordinate the solar generation, energy storage and transmission components. These so-called ''Smart Grid'' technologies include: advanced metering, meter database automation, quasi-real time bi-directional communications between customers and producers, direct load control, central distributed generation control and intelligent appliances.

Customers will increasingly play a larger role in addressing the challenges of our energy future. Smart Grid technologies provide both customers and utilities with the tools to better manage the production, storage, delivery and use of electricity. In so doing, the Smart Grid changes the basic premise of electricity providers, transforming utilities from providers of am energy commodity to enablers of energy transactions. Under such circumstances, regulations to decouple commodity energy supply from utility revenue recovery will be an integral part of the development of the Solar Grand Plan.

Innovative rate incentives could be developed to make effective use of a nationwide Smart Grid. For example, owners of plug-in hybrid electric vehicles in New Jersey might be convinced to lend use of their cars' batteries to store solar energy produced in Arizona. Continued support of the National Laboratories for development and testing of Smart Grid technologies and development of transaction manager grid control algorithms will enable the Solar Grand Plan technologies to become reality.

Other Considerations

The Solar Grand Plan requires the commitment of fairly large tracts of land for the installation of solar generation, energy storage and transmission. Transmission and CAES storage facilities would be spread out across the country, mitigating their environmental impact on any particular region. However, the solar generation component is expected to be concentrated in the southwestern U.S. to tap promising solar energy resources in Arizona, New Mexico, Nevada, Texas and California. This could create concerns about the environmental impact of large sections of land being effectively covered with solar collectors. Placing a large percentage of our nation's solar generation assets in one geographic area also makes them more susceptible to damage from a single weather related event. Distributing solar generation systems over a wider area may reduce that risk of damage.

Funding should be made available for evaluation of optimum solar generation area coverage factors, heat island creation, environmental mitigation, wildlife habitat impacts and beneficial land uses in harmony with solar collectors. We also would need to consider the societal impact of bringing a significant number of solar equipment installers to live and work in currently uninhabited areas of the desert south-west.

While the Solar Grand Plan envisions that most of our sustainable energy resources will be solar, the system would accommodate any generation resource—including our existing coal and gas fired power plants and nuclear facilities. As we transition to a new solar-based energy infrastructure, we must make accommodations for linking these existing resources into the national transmission backbone and incorporating their output into our energy storage plans.

Financial Support

As with all energy project proposals, the Solar Grand Plan has a price. An estimated subsidy of $420 billion would be needed to support development of plan components from 2009 to 2020. After 2020, the plan should be financially self-supporting as the cost of solar power with storage drops below the price of energy that could be generated from proposed traditional power plants. At that time, the Solar Grand Plan infrastructure would provide an economic alternative to construction of those new plants, and funds for continued expansion of the system would come from the sources traditionally used for new power plants today.

A commitment to fund the solar generation portion of the Solar Grand Plan would encourage solar manufacturers to invest in new production factories in the United States. We will not fully reap the economic benefits of solar power development or achieve national energy security until the manufacturing of our basic energy Solar Grand Plan components occurs within this country. We will also not take full advantage of reducing our energy related expenses overseas if we are purchasing solar products produced in other nations.

Conclusion

The Solar Grand Plan is proposed to demonstrate that there is at least one feasible, affordable, realistic plan based on proven existing technologies that can transition our fossil and nuclear energy based electric energy production infrastructure into a sustainable energy production infrastructure. It is instructive that the Interstate Highway System had an initial cost of $425 billion in 2006 dollars, very close to the $420 billion of subsidies that would be needed to bring the Solar Grand Plan into reality.

The Solar Grand Plan strengthens our energy security by effectively storing electrical energy underground, as we have done with fuels at the Strategic Petroleum Reserve and at natural gas storage facilities. Energy security is further enhanced by geographic dispersion of the energy storage facilities and through redundancy in the transmission backbone right-of-way alignments. Just as the construction materials and route alignments evolved during the 35-year construction of the Interstate Highway System, the Solar Grand Plan will benefit from advancements in technology and regulations during its development. And just as it took the political will of Congress with strong support from the states to make the Interstate Highway System a reality, strong leadership, vision and the will of Congress in partnership with the states will be essential to implementing a long-term, reliable, sustainable, secure energy solution for our nation.

Chairman Lampson, Vice Chair Giffords, Ranking Member Inglis and distinguished Members of the Committee and staff, I want to thank your for this opportunity to address the Solar Grand Plan and for your dedication to finding solutions to the energy challenges facing our future.

Biography for Thomas N. Hansen

Education

Lehigh University, BSEE, 1971

- Major—Electrical Engineering/Computers, Hardware and Software Design
- Minor—Physics

Stanford University, MSEE, 1972

- Major—Electrical Engineering/Computers, Hardware and Software Design, Laser Design, Inertial Navigation Design
- Minor—Geophysics

Work Experience

Tucson Electric Power Company—1992–Present

Vice President, Power Production—1992–1994

Vice President, Technical Adviser—1994–2006

Vice President, Environmental Services, Conservation and Renewable Energy—2006–Present

Guiding the development of the Renewable Generation Portfolio of Tucson Electric Power, including 11 MW of renewable generation capacity of which six MW is solar. Currently Arizona's largest single utility renewable generation fleet.

Alamito Company/Century Power—1984–1992

Superintendent of Operations, Springerville Generating Station—1984–1986

Plant Manager, Springerville Generating Station—1987–1988

Vice President, Operations, Springerville Generating Station—1989–1992

Bechtel Power Corporation—1972–1984

Senior Field Engineer, Navajo Generating Station—1972–1976

Start-Up Engineer, Cholla Generating Station—1977

Assistant Electrical Superintendent, Coronado Generating Station—1976–1980

Electrical Superintendent, Palo Verde Nuclear Generating Station—1981

Project Field Engineer, Springerville Generating Station—1981–1984

Career includes the design, construction, operation or management of over 10,000 MW of electric generation capacity in the western United States.

Affiliations Present/(Past)

EEI Renewable Energy Working Group—Chair, 2 yrs.

Arizona Governor's Renewable Energy Working Group—1 yr.

International Energy Agency Solar Energy Task 8 Committee Member and U.S. representative—2 yrs.

(WGA Clean and Diversified Energy Advisory Committee Solar Subcommittee Member —1 yr.)

(Utility Photovoltaic Group Board—Member, 3 yrs.)

Professional Licenses

Registered Mechanical Engineer—Arizona

Registered Electrical Engineer—Arizona, California

Publications

"Energy Pay-Back and Life Cycle CO_2 Emissions of the BOS in an Optimized 3.5 MW PV Installation." J.E. Mason, V.M. Fthenakis, T. Hansen and H.C. Kim. *Progress in Photovoltaics,* May 6, 2005.

''Photovoltaic Pourer Plant Experience at Tucson Electric Power.'' Larry Moore, Hal Post, Tom Hansen and Terry Mysak. *2005 ASME International Mechanical Engineering Congress,* November 2005. IMECE2005–82328.

''Environmental Portfolio Standard Meeting Solar Electric Generation Goals: The Utility View.'' Presented October 25, 2002 in Phoenix, Arizona at the Arizona Corporation Commission.

''More Solar for Less $.'' Presented October 3, 2003 in Scottsdale, Arizona at UPEX 2003: Solar Power Experience Conference.

''The Systems Driven Approach to Solar Energy: A Real World Experience.'' Presented October 15, 2003 in Albuquerque, New Mexico at the Sandia Laboratories Solar Power Conference.

''The Promise of Utility Scale Solar Photovoltaic (PV) Distributed Generation.'' Presented March 2, 2004 in Las Vegas, Nevada at the Power-Gen Renewable Energy 2004 Conference.

''Utility Scale Photovoltaic (USSPV) Distributed Generation.'' Presented April 5, 2004 in Phoenix, Arizona at the Arizona Corporation Commission.

''Springerville Generating Station Solar System: A Case Study.'' Presented October 19, 2004 in San Francisco, California at the Solar 2004 Conference.

''Utility Scale Photovoltaic Generation: Broccoli for Utilities—A New Generation Financing Paradigm.'' Presented October 27, 2005 in Phoenix, Arizona at the Arizona Corporation Commission.

''Utility Scale Photovoltaic Generation: A New Opportunity.'' Presented November 8, 2005 in Denver, Colorado at the Rocky Mountain Electric League Fall Conference.

Ms. Giffords. Thank you, Mr. Hansen.

Ms. Maracas.

STATEMENT OF MS. KATE MARACAS, VICE PRESIDENT, ARIZONA OPERATIONS, ABENGOA SOLAR INC.

Ms. Maracas. Thank you. Chairwoman Giffords, Members of the Committee, and staff, I, too, would like to express my appreciation for the opportunity to talk about this very important subject today.

Thank you.

I am the Vice President of Arizona Operations for Abengoa Solar, and we are a very large company based in Madrid, Spain. Abengoa employs over 23,000 people worldwide, and we have presence in more than 70 countries. Right now, we have about 40 people in the U.S. and Spain, who are dedicated to improving the technology and developing solar technology in the sunny southwestern states.

In December of 2007, last year, the U.S. Department of Energy selected us for three R&D projects aimed at improving solar parabolic trough technology, which you saw in Mr. Mehos' presentation. And recently, as we have already talked a little bit about, we have announced an agreement with Arizona Public Service to build, own and operate the 280 megawatt CSP, or concentrating solar plant called Solana.

APS will purchase all of the output of the plant, and I think as Congressman Mitchell already mentioned, if the plant were in operation today it would be the very largest in the world.

So, with over 500 megawatts of large-scale solar power plants in operation, development and construction in the U.S., Spain, Algeria, and Morocco, I think our company is notably one of the largest providers, leading providers of large-scale solar technology today. And, with that position in mind, we are grateful to be part of this important dialogue and discussing the role that CSP and other large-scale solar technologies can play in helping build our nation's energy resource portfolio.

We will also be talking about the opportunities for removal of obstacles or barriers that could get in our way, and could otherwise prevent us from leveraging this very abundant resource, which as we heard comes in buckets from the sky. I like that analogy, I think it is a good one.

I have been asked to address a few topics today, and they include, one, the efficacy of large-scale solar power as a significant component of the U.S. generation fleet, and barrier reduction opportunities for achieving this potential. Two, the near and long-term economic impacts of large-scale solar deployment. And third, the role of government in advancing solar thermal technologies.

In the interest of time and efficiency, I will probably skip the third one, because I really cannot add anything to what Mr. Mehos has already commented, and I certainly concur with his remarks.

On the subject of large-scale solar generation as a viable option for providing significant contributions to our power needs, my view is, certainly, that large-scale power facilities not only have the potential to become a leading part of our national resource portfolio, they are also among the smartest options that we can exercise, particularly, in a business-wise context.

Further, I see today's family of CSP technologies as an important mainstream option for utility resource plans, and I will explain the reasons for those thoughts momentarily. Just one minute I will spend, before I get there, though, on the distinction in technologies. Mr. Mehos already pointed out some of the few, but the family of solar and thermal CSP technologies is growing rapidly and there is an increasing number of technologies that are becoming part of the solar thermal or CSP family. These technologies are advancing rapidly in the marketplace, but there are two basic categories of technology that I would like to distinguish.

One is the category of photovoltaic or PV technologies, and those are the technologies that convert the sun's energy directly into electricity by virtue of a photo electric reaction that occurs on a semiconducting material. And, when a concentrating mechanism such as a lens is used on conjunction with those PV cells then we have a large-scale technology known as high concentration photovoltaics or ACPV, and this becomes a member of the utilityscale solar family, as I mentioned earlier, and you saw an illustration of one of those technologies on Mr. Mehos' slide.

But, the solar thermal category, which is kind of the work horse of the large-scale family, is a bit different, in that it uses the sun's heat to produce steam, which in turn becomes the working agent in a conventional ranking cycle, the very familiar thermodynamic process that converts heat to energy in a common steam plant. So, this is very familiar, tried and true, mature technology.

The significant difference between our kind of solar steam plant and natural goal or coal steam plants is that there is no fossil fuel combustion or associated carbon emission to use in creating the mechanical energy that in turn spins the turbine, and then transfers mechanical energy to an electric generator.

So, most of my remarks today are with the large-scale solar technologies in mind.

And then, returning to my comment that CSP is a business-wise decision, I can offer that at Abengoa Solar we talk with many, many utilities in our sunny southwestern and western states, and they are beginning to articulate large-scale solar in a different way. It is no longer just something that we have to do for compliance anymore. It is no longer something that our utility colleagues talk about as an R&D endeavor or an experiment, it is something that they consider a wise part of the their future resource planning options.

When we talk about advanced coal technologies, which are not terribly mature yet today, and what are the other options that could compete as we are thinking about rapidly-growing demand in service areas, you probably will hear from Ms. Lockwood today about the rapid growth that APS is experiencing in our state.

When you look at natural gas volatility risks, and the increasing likelihood of some kind of carbon regulation or carbon tax, then utilities are really beginning to think of it differently. Although, as Tom says, everything has a price, there is a slight premium today above conventional generation costs for CSP or large-scale solar generated electricity, but that cost gap is closing and I think Mark's slides illustrated that very well. As natural gas prices rise, as other costs go up, and as carbon regulation becomes more imminent the gap is closing. Our costs will come down.

Utilities are increasingly viewing CSP as a wise bet against fuel price volatility and open-ended carbon liability.

APS, our first large-scale CSP customer in the U.S., has, in fact, been very, very forward thinking about the role that CSP will play in their future resource portfolio, and I will let Barbara talk about that in just a little bit. But, I will say that APS has really been a leader among a group of very proactive utilities who are thinking about this in a very different way today.

And, the final portion of this first topic that I have been asked to address relates to those barriers that could stand in the way of large-scale solar deployment, and I will just be very simplistic about this. In my mind, there is no barrier whatsoever related to technology. Yes, indeed, there is room for improvements in cost, and performance, and efficiency, and R&D should continue, but in terms of will technology be the barrier that prevents us from going forward, it most definitely will not.

In my view, the greatest barrier to increased deployment of solar generating facilities is, indeed, political rather technical, and you know where I am going with this, I am going to be talking about the ITC, the lack of an enduring tax credit, the 30 percent solar investment tax credits that we have talked about briefly already, is really the biggest hindrance that we see today in large-scale solar.

The ITC has been in place since the passage of the Energy Policy Act in 2005, but it has, since its enactment it has really just been kept on life support with one or two-year extensions at a time.

As Mark said, these one or two-year extensions do large-scale projects really no good, in fact, they do a disservice, because it prevents us from developing a longer extension through Congress that could help create the certainties that capital markets need to lend money on these large projects. The duration of one or two years, which is shorter than the project development time for a large-scale solar plant, means that we really must see a long-term extension, at least eight years of the ITC, in order for this industry to move forward.

As Congressman Mitchell mentioned, we have been very candid about the fact that the Solana Generating Station in the Gila Bend area of Arizona cannot happen without it, without the eight-year extension.

So, for these reasons I guess I will just say, ITC, ITC, ITC over and over again, and I would urge Congress to extend this important measure for an eight-year period, through bipartisan support of the Renewable Energy and Energy Conservation Tax Act of 2008 which passed through the House last month.

On the subject of near and long-term economic impacts of large-scale solar deployment, I can draw observations from a very large body of credible research that has been done over the last several years on this topic, and I will speak as a member of the Western Governors' Solar Task Force—or the Western Governors' Association Solar Task Force, as a member of that group I participated in a comprehensive effort to analyze the role that solar energy could play in helping the Governors meet an ambitious goal of deploying 30,000 megawatts of clean energy in their 19 states by 2015.

So, our task was to understand what is the actual resource potential, what is the market potential, how does that match up with demand for energy, what is the industry's capability to gear up and build projects and deliver energy through solar resources, and also to understand the barriers to deployment, and then finally, what will this effort pay back in terms of economic benefits.

On the topic of economic impact, we examined over a dozen economic studies that had been done since 2004, all by credible investigators, State governments, national labs, universities. In fact, three of those studies were supported by Mark's institution, the National Renewable Energy Laboratory, and these three looked specifically at CSP, Concentrating Solar Power plants, and in a variety of different scenarios and assumptions about different growth patterns, and sizes of deployments and so on, they looked at what would be the impacts in terms of private investment in the state, permanent and temporary job creation, indirect and direct effects, and so on.

And, we convened—these studies were conducted for Nevada, for New Mexico, and for southern California, and so the assumptions were all different, as I mentioned, different sizes of economies, and different scales and so on, in order to just understand, in a general sense, what does CSP or large-scale solar deployment really mean, in terms of just kind of generalization across the board, what does one gigawatt, 1,000 megawatts of CSP actually do to our economies. So, we convened an expert panel of economists to generalize these impacts, and our findings were that for every gigawatt of CSP added to the state's economy the deployment would yield \$3 to \$4 billion of private investment in the state, 3,400 to 5,000 construction jobs, and up to 200 permanent solar plant jobs, many of those in rural areas where we typically are trying to attract economic development, a \$1.3 to \$1.9 billion, 30-year increase in tax revenues, and that is after any tax incentives or other incentives are given to the project developers, and between \$4 and \$5 billion in increased gross State product. And so, a general rule of thumb for one gigawatt of CSP. These are enormous positive impacts that can occur.

Clearly, the findings indicated that the broader incorporation of large-scale solar plants into the U.S. generation fleet, not only produces those economic impacts, but also the benefits of sustainability and energy independence, as Mr. Hansen spoke about.

Finally, on the role of government advancing solar technologies, as I said, I will defer. I think those points have already been made very well. I will say that the thing that government can do for this technology, in addition to what Mr. Mehos mentioned, and issues that will help us site and facilitate permitting, solve transmission congestion problems and so on, we have to help CSP learn to walk on its own, and, in fact, we very much hope it learns to run.

Thank you very much.

[The prepared statement of Ms. Maracas follows:]

Prepared Statement of Kate Maracas

Mr. Chairman, Vice Chairman Giffords, Members of the Committee, thank you for the opportunity to testify today. I am the Vice President of Arizona Operations for Abengoa Solar Inc., a U.S. division of Abengoa, which is based in Madrid, Spain. Abengoa employs over 23,000 people worldwide, with presence in more than 70 countries. Abengoa Solar has a team of approximately 40 people in the United States and Spain dedicated to researching, developing and improving solar technologies. In December 2007, the U.S. Department of Energy selected Abengoa Solar for three research and development projects to improve solar parabolic trough technology. And recently, we announced an agreement with Arizona Public Service to build, own and operate a 280 Megawatt (MW) Concentrating Solar Power, or ''CSP'' plant in western Arizona. APS will purchase all of the output of the plant, known as the Solana Generating Station. If in operation today, Solana would be the largest solar power plant in the world.

With over 500 MW of large-scale solar power plants in operation, development, and construction stages in the U.S., Spain, Morocco, and Algeria, Abengoa Solar is notably one of the world's leading providers of large-scale solar technology solutions today. With that position in mind, we are especially grateful for the opportunity to be a part of this important dialogue about the role that CSP and other large scale solar technologies can play in our nation's energy resource portfolio, and the opportunities for removing obstacles that could prevent us from leveraging our very abundant and sustainable solar resource.

I have been asked to address a few topics today, and they include:

(1) The efficacy of large-scale solar power as a significant component of the U.S. generation fleet, and barrier reduction opportunities for achieving this potential;
(2) Near- and long-term economic impacts of large-scale solar deployments; and
(3) The role of government in advancing solar thermal technologies.

I will attempt to address these topics, in that same order.

On the subject of large-scale solar generation as a viable option for providing significant contributions to our nation's power needs, my view is that large-scale solar power facilities not only have the potential to become a meaningful part of our national resource portfolio; they are also among the smartest options we can exercise—particularly in a business-wise context. Further, I see today's family of CSP technologies as an important ''mainstream'' option for utility resource plans. I will explain the reasons for those thoughts momentarily, and before I do, a brief discussion about the distinction between large-scale solar generation and CSP in particular is worthwhile.

The family of solar thermal and CSP technologies is growing rapidly. An increasing number of technology approaches to solar thermal generation is advancing in the market place. I would like to clarify that there are two very basic categories of solar electricity generation. One is the category of photovoltaic, or ''PV'' technologies —those that convert the sun's energy directly to electricity by virtue of a photo-electric reaction that occurs on a semi-conducting material. When a concentrating mechanism such as a lens is used in conjunction with PV cells, the technology is known as High Concentration Photovoltaics, or ''HCPV.'' Because the lenses add great efficiency to the PV cells' production capacity, HCPV is currently being developed as a utility-scale solar option.

The solar thermal category is a bit different, in that it uses the sun's heat to produce steam, which in turn becomes the working agent in a conventional Rankine Cycle—the very familiar thermodynamic process that converts heat to energy in a common steam power plant. The significant difference is that a solar thermal plant requires no fossil fuel combustion or associated carbon emissions to create the mechanical energy that spins a turbine, which in turn transfers mechanical energy to an electric generator.

Most of my remarks today contemplate thermal CSP technologies, although Abengoa Solar also views HCPV as a very promising technology in the near horizon.

Returning to my comment that CSP is a ''business-wise'' decision, I can offer that Abengoa Solar Inc. holds discussions with many utilities in our sunny western and southwestern states, and an increasing number of our utility contacts articulate that they no longer view CSP as *just* an option for Renewable Portfolio Standard (RPS) compliance, or as an experimental R&D endeavor. Rather, our utility colleagues consider their future resource planning options in the context of advanced coal technology and emission constraints, natural gas price volatility risks, and the increasing likelihood of carbon emission costs in the form of externalities or direct taxation. Although there is a slight premium today above conventional generation costs for CSP-generated electricity, the cost gap is closing as fossil fuel prices increase and carbon regulation becomes more imminent. With today's promise of dispatchable solar plants made available through advanced commercialization of Solar Thermal Energy Storage (TES) technology, utilities increasingly view CSP as a wise bet against fuel price volatility and open-ended carbon liability.

Arizona Public Service, our first large-scale CSP customer in the U.S., has in fact been very forward thinking about the role of CSP in their future resource portfolio. APS is a leader among a group of proactive utilities in our nation who very definitely view CSP as a viable part of a low-risk resource portfolio, and as a mainstream element of their growing generation fleet.

The final portion of this topic that I have been asked to address relates to those barriers that may stand in the way of large-scale solar deployments. There is no question in my mind that technology is not a barrier. While there is room for cost and performance improvements that will occur with technology advancements, economies of scale, repetition and associated learning curve improvements, the greatest barrier to increased deployment of solar generating facilities is indeed political rather than technical. While federal support of R&D must continue, the single most significant hindrance to broader deployments of CSP facilities in the U.S. is the lack of an enduring tax credit which is essential to the financial viability of CSP installations today. The 30 percent federal Investment Tax Credit, or ''ITC,'' has been in place since passage of the *Energy Policy Act 2005*. But since its enactment it has been kept on life support with one- or two-year reauthorizations at a time. The short lifespan of the ITC does not stimulate the deployment of large, capitalintense solar generating stations, which require three to four years to build. Further, the large institutional entities required to provide construction and operating capital for these projects cannot operate with the uncertainty of an expiring tax credit whose duration is shorter than a project development period.

In summary, are there technology improvements to be achieved for large-scale solar through R&D? Absolutely. Are the barriers to meeting more of our nation's energy needs through solar energy production related to technology? Absolutely not. The single most important barrier to achieving our solar potential is the lack of a policy framework that is sufficiently robust to stimulate solar deployments in a meaningful way. We, our industry colleagues, and our consumers urge Congress to extend the federal ITC for an eight year period through bipartisan support of the *Renewable Energy and Energy Conservation Tax Act of 2008* that passed in the House last month.

On the subject of near- and long-term economic impacts of large-scale solar deployments, I can draw observations from a large body of credible research that has been done over the last several years. As a member of the Western Governors Association's Solar Task Force, I participated in a comprehensive effort to analyze the role that solar energy could play in helping the Governors meet their goal of deploying 30,000 MW of clean energy in their 19 states by the year 2015. Our task was to understand the resource potential, the market potential, the industry's capacity, the barriers to deployment, and the economic impacts that would result. On the latter topic, we examined over a dozen economic studies conducted since 2004 by credible investigators such as universities, national laboratories, and State governments. In fact, three of those studies, supported by the National Renewable Energy Laboratory (NREL) examined the economic impacts that could be expected as a result of increased deployment of CSP plants in particular. The studies contemplated a variety of CSP plant growth and scale scenarios, and the changes to be expected in terms of job creation, net Treasury gains, Gross State Product, and private investment.

We convened an expert panel of economists to generalize these impacts across different State economies, and across different assumptions used among the studies.

Our findings were that for every one Gigawatt (GW) of CSP added to a state's economy, the deployment would yield:[1]

- $3–$4 billion private investment in state;
- 3,400–5,000 construction jobs; up to 200 permanent solar plant jobs, many in rural areas;
- $1.3–$1.9 billion 30-yr. increase in State tax revenues; and
- $4–$5 billion increase in Gross State Product.

Those figures represent net effects, even after any tax credits or economic incentives are utilized to stimulate industry development. Clearly, the findings show that broader incorporation of large-scale solar plants into the U.S. generation fleet not only produces the benefits of sustainability and energy independence, it also pays back in very significant, positive economic impacts.

Finally, on the role of government in advancing solar thermal technologies, it is clear that the private sector cannot achieve a ''Grand Solar Plan'' alone. The market penetration of any new technology, product, or service traditionally follows a pattern of growth in market adoption, followed by declining prices and higher margins that result from economies of scale. Large-scale solar generation is no different in that regard. What is different, however, is that the capital commitments required to bring large-scale solar plants to market are very large, and the risk of investing in such markets with the hope that demand will follow is too high for private sector entities to bear alone. This condition describes the very traditional role that government has played in numerous examples of infrastructure development and market stimulation actions.

The government's role in solar power thus far has been both push and pull. By that I imply that the creation of demand for clean solar energy in the market place must come from both mandates and incentives. Twenty six states, including Arizona, now have Renewable Portfolio Standards that require increasing portions of delivered electricity to be derived from renewable energy resources. The RPS frameworks are a very good start, but only speak to half of the push-pull equation. Governments must also step up to the plate to incentivize market activity, and so I repeat here that a vitally important role for the Federal Government will be to extend the ITC for eight years so that large solar power plants can be financed and be economically viable. Recalling my comparison to other new technologies, products, and services in the marketplace, CSP will also grow up and learn to walk on its own.

On a final note, I will comment that we are very pleased to see the serious commitment to solar energy R&D that both the President and Congress have demonstrated in recent years. While I noted earlier that technology itself is not a barrier to large-scale solar power production, the efficiency and performance improvements that will be accomplished through R&D will continue to be an important part of ongoing cost reductions that will help large-scale solar generation to walk on its own. In fact, we hope it learns to run.

Thank you very much for the opportunity to share our perspective on this important topic.

Biography for Kate Maracas

Kate Maracas is the Vice President of Abengoa Solar's Arizona Operations, where she focuses on infrastructure development to support the siting of Concentrating Solar Power (CSP) plants within Arizona. She has more than 25 years of energy industry and consulting experience in areas including engineering, environmental management, and renewable energy. Her career emphasis over the last several years has been in large scale solar generation. Kate actively participated with the Western Governors' Association Solar Task Force in developing its report and recommendations for solar market expansion in the western region, and continues to work on legislative and policy efforts to advance CSP deployment. Kate holds a Graduate Certificate in International Business from the Thunderbird Graduate School of International Management, and a Bachelor of Science Degree in Electrical Engineering from Arizona State University. Ms. Maracas currently serves as an appointee of Governor Janet Napolitano on Arizona's Solar Energy Advisory Council, and chairs the CSP committee of the Governor's Council.

Ms. Giffords. Thank you, Ms. Maracas.

Ms. Rauluk, please.

STATEMENT OF MS. VALERIE RAULUK, FOUNDER AND CEO, VENTURE CATALYST INC.

Ms. RAULUK. Thank you, Madam Chair, Chairman Gordon, and distinguished Members of the Committee, Representatives Lipinski, Matheson, Mitchell, and Ranking Minority Leader Hall. I would also like to thank my colleagues regarding the insights I will share with you. Please note that Venture Catalyst is solely responsible for the opinions expressed. Thanks to Arizona Research Institute for Solar Energy, AzRISE, the University of Arizona, Joe Simmons and Ardeth Barnhardt, my colleagues at Sun Edison, Jigar Shah, Howard Green, Colin Murchie, and my colleagues at Raytheon Missile Systems, John Waszczak and Thomas Olden.

Venture Catalyst has been engaged in commercialization of solar energy technology for the last ten years, primarily, under contract with the U.S. Department of Energy, Sandia National Labs, and the National Renewable Energy Laboratory. We have also worked with regional and national solar energy companies in thermal and photovoltaic technologies, from residential to utility-scale markets.

Although fundamentally agnostic when it comes to solar technologies, I will address specific preferences based on risk factors and maximized benefits.

Here is what I hope to cover, opportunities, obstacles, critical factors in the first 10 years.

The ''Grand Solar Plan'' is possible with the right investment structure and technology portfolio. The path to progress for achieving that goal is a going-forward focus on Distributed Solar Photovoltaic, DGPV, I hope to use that, DGPV, throughout here.

Given the dominance of PV in the ''Grand Solar Plan'' and the 3,000 megawatts of solar thermal projects in the pipeline, a similar commitment to PV should be made. At present, 800 megawatts of PV have been cumulatively developed.

But most importantly, the DGPV focus offers the greatest flexibility for resolving the biggest obstacle to achieving the plan financing. Strategically targeted DGPV development offers all of the investor groups the best return on their investment, especially, the rate payers, and I am going to talk about that in a little more detail in a minute.

A couple critical factors for success on the path to progress. First of all, productively framing the concept of utility-scale solar. The second, continuing technical improvements, and third, structuring the investment for fairness and maximized benefit.

I want to talk a little bit about the different technologies and the different formats. You have already seen the difference between PV and solar thermal, I want to go into that.

There is two major development formats for solar energy. One is the central station format, and the other is the distributed format.

This is an example of solar thermal projects, which because of economies of scale limitations are to date developed in large-scale central station format.

The title of this image is ''Perfection.'' This is one of the many formats that solar photovoltaic projects can take, a utility-scale project in a distributed generation DG format. It does not require transmission and the associated costs considered connected to the distribution part of the electric power grid. Strategically located PV installations within the distribution grid can provide benefits above and beyond the value of clean green peak power, benefits that can extend the life of power infrastructure and increase the reliability of the local grid.

This is another picture of that same installation of an 8.4 megawatt in southern Colorado, and you can see the substation in the background there.

Here is another way to deploy DGPV, and it has actually been the major development mode for PV to date, and that is roof-top installation.

I would like to offer the following definition of utility-scale solar energy and call out the key factors, large scale, lower cost, higher reliability, and benefits across customer classes.

Here are a couple examples, two to 20 megawatt solar farms strategically located in the low pockets, one to five megawatt solar farms on the roof tops of schools, reducing school energy expenditures, and putting solar energy into the grid in the summertime for the rest of the community, and also, of course, the one to five megawatts in commercial government installations, similar to what I showed you before.

Critical factor 2, solar energy technology is ready right now. You have heard that from several of the speakers. PV, in particular, has, and continues to achieve, incremental improvements across the entire value chain. Additionally, major advances are coming out of Arizona laboratories in the next three to 10-year term, specifically, in the area of PV, concentrated PV, or CPV, and high-concentrated PV, or HCPV. A solar focus on the PV family of technologies will help harvest additional economic benefits for Arizona and Arizona rate payers, and you will have to pardon my specific interest in Arizona.

A major obstacle to PV deployment has been the conflict with preserving utility revenues. By means of both product design and regulatory rules and policy, the industry is actively addressing this conflict. A commitment to large-scale DGPV programs would accelerate the resolution.

Financing, financing, as I have said, is a major obstacle to making this happen.

We are asking the Federal Government, and local government, private capital, and, most importantly, the rate payers, from whence the utility renewable energy investment dollars comes from, are all the people involved in this transaction.

We are asking rate payers, I think this is the most important thing that you need to think about as we structure this, we are asking rate payers to contribute to solar energy investment at the same time they are being asked to pay for fossil fuel increases. By developing many solar projects across numerous communities, more rate payers will experience the direct benefits of economic development and increased power security and reliability. PV, CPV, HCPV, because of greater scalability, flexibility, and modularity, can be deployed more easily across a range of situations.

Regulatory obstacles, we know what has to be done, and have developed and implemented best practices in key regional U.S. markets. We need to nationalize those practices for a more rational and effective market. I would concur with many of the things that people have said already, the ITC is very important, and nationalizing some of these best practices is very important as well.

Incentive design, in general we need to reallocate our incentives away from fossil fuel and to solar energy, and we know there are some best practices on how to design that.

So, for the first 10 years, my recommendation is reallocate investment to the highest benefit, lowest-risk installations, and that is DGPV, and to aggregate smaller projects in a systematic program to reduce risk while delivering the cost of scale. And, I believe that this will effectively stretch public investment, and it will reduce performance and project risk.

In the second five years, I would see that we would integrate higher-volume concentrated PV and high-concentrated PV as the cost effectiveness and reliability of those technologies come on line, and begin a second phase of solar thermal development if feasible or necessary.

So, in conclusion, the path to progress for the Grand Solar Plan, make balanced, judicious, timely technology choices, structure the investment for fairness and efficiency across all of the investment groups, and match the 3,000 megawatt solar thermal projects in process with 3,000 megawatts of photovoltaic projects.

Thank you for this opportunity to testify, and thank you for your leadership in this critical issue, and especially, Chair, Congresswoman Giffords, I really appreciate your leadership here. It has made a very big difference in Arizona.

Thank you.

[The prepared statement of Ms. Rauluk follows:]

Prepared Statement of Valerie Rauluk

Madame Chairman, Members of the Committee, distinguished guests I am honored to offer my testimony concerning Utility Scale Solar Power. My comments will address the following:

1. Grand Solar Plan as a viable option. The technical & regulatory obstacles.
2. Current solar energy market and expected changes over the next 10 years.
3. Current regulatory environment and incentive structures conducive to large scale solar development & recommended improvements.
4. Distributed PV and concentrating PV compared with solar thermal technology. Areas of government research that can play a critical role not met by private sector. Other recommendations and priorities.

1. The Grand Solar Plan as a Viable Option. The Technical & Regulatory Obstacles

The Grand Solar Plan calls out a vision of nearly 70 percent of our electricity generation from solar energy by 2050. It also calls for a technology mix of five times as much solar photovoltaic (''PV'') as solar thermal. This vision is highly probable, with the right development framework and investment incentives.

Additionally, the Grand Solar Plan calls out for a development format of large-scale remote solar energy generation, compressed air storage, and direct current transmission. My colleagues at the University of Arizona have convinced me that compressed air storage and direct current transmission are more than science fiction, although there is much that needs to be assessed for both approaches to be viable. It is important to note that designing, financing and implementing a large-scale adoption of such strategies is no minor feat. As such, I would see, from my experience and understanding of technology adoption cycles, that such approaches will not be available for commercial adoption for 10 or more years. Other witnesses could clarify the risks, timing and benefits better than I. Beginning the process of assessing and designing such approaches is useful, but I would caution that we focus on the approaches that can deliver large amounts of market driven solar generation into the mix quickly, with the lowest risk and the greatest benefits.

As you will see, my comments focus on the first ten years of a Grand Solar Plan. The first steps will be difficult. Large amounts of investment capital from public and private sectors will be needed. And the skeptics concerning solar energy and its primary role in the greening and cleaning of our energy system will be numerous and loud. That is why in the first years, we should focus on efforts that lower risk and maximize the benefits.

In addressing the technical and regulatory obstacles to achieving the Grand Plan the following critical factors will be addressed.

- Critical Factor #1—Productively framing the definition of ''utility scale solar'' and supporting with regulatory requirements.
- Critical Factor #2—Technology improvements, including improvements in business model.
- Critical Factor #3—Effectively structuring the multi-billion dollar investment to be made by rate-payers, investors and the government.

Critical Factor #1—Productively Framing the Definition of ''Utility Scale Solar'' and Supporting with Regulatory Requirements

Although the Grand Solar Plan does not make specific recommendations regarding the development format, it seems to imply, with the recommendation for large scale storage and specialty transmission, that solar energy should be developed under the model of the last 50 years: large scale, remotely located, dependent on extensive transmission for delivery to consumers. This is commonly referred to as the ''central station'' model.

There is a more market driven way to develop solar energy and the successes of the last few years highlight the approach. That approach consists of smaller generation facilities, on otherwise unused real estate (roof-tops and sites of 10 to 500 acres of land, two percent to 80 percent of a square mile), located near the load demand, and dispersed throughout many communities. That approach is called distributed generation or ''DG.''

DG is not only a path of more rapid, less risky development, it is also the path for a more robust power network. There is an inherent resiliency in networked systems where resources are at the point of use (like the Internet) instead of a hub and spoke development (like the land-line telephone system). This resiliency can be enhanced with the addition of small scale, on-site intelligent controls and storage, increasing reliability and dependability and improving the fit between resource generation and needs across the local grid. When developed in a strategic and coordinated fashion, DG can delay or eliminate the need for distribution and transmission infrastructure investment.

Although DG includes very small generating systems, much larger DG systems (up to 20MW in a single location) can be clearly characterized as ''utility scale'' as can a systematic aggregation of many smaller generators. Utility scale can be more productively thought of as any project/program offering high volume, lower-cost, reliable, and dependable renewable energy for 20 years plus at fixed prices for large numbers of customers. Utilizing this more expansive definition of utility scale offers more options for maximizing solar energy deployment at the best cost-benefit tradeoff, starting now.

Examples are:

- 2 to 20MW solar farms, strategically located in load pockets to strengthen the grid and increase community energy security in case of transmission failure.
- 1 to 5 MW solar farms on the roofs of our schools, reducing school budget exposure to volatile and rising energy prices for 20 years and pumping solar power into the grid for community use during the summer days when community demand is most pressing.
- 100kW to multiple megawatts on commercial, government, industrial sites/buildings.

In Arizona alone, an immediate potential of multiple gigawatts of solar energy systems are available. With expected cost reductions, 65 GW of solar energy could be developed in the U.S. over the next ten years (U.S. Department of Energy. Solar America Initiative, *http://www1.eere.energy.gov/solar/solar¥america/*).

Central station development has its attractions. It feeds into the 'bigger is better' syndrome. Bigger means more attention. Bigger means larger development fees. Usually bigger means cheaper. But what a great deal of research has shown, is that bigger, especially when it comes to power plants, can often be riskier: longer construction periods, higher financing cost, longer delays before a system is producing and selling energy to end-users, to name just a few.

Bigger also usually means more remotely located from where the consumers are, a distance that results in additional costs for transmission: wheeling charges, transmission investment and public approval (nobody seems to want transmission lines in their backyards), and transmission losses. And last, but not least the security exposure of having a critical resource like power, vulnerable to hundreds of miles of difficult to protect delivery infrastructure.

Removing the Obstacles Since nearly all of our existing power generation is central station, and a considerable amount of central station solar power is in the early stages of development, our focus going forward should be to diversify our resource portfolio and focus on solar DG installations.

Regulatory Regulatory obstacles to this path are fairly straightforward and in fact, many states have established law, policy and procedure to remove them. That is how 300MW of solar energy got developed last year. But the patchwork has prevented a truly vibrant and efficient market for solar energy. Efforts at the federal level to establish the following best practices will accelerate the development of solar energy.

Level the Playing Field for Incentives, Subsidies and Financing Establish incentives at the federal level that match the incentives given to fossil fuels. Structure for rapid and long-term deployments with declining levels of support to encourage systematic and focused cost reduction across the whole value chain. Reward system performance and support system diversity,

Net Metering Require full retail value for all solar energy produced by customers without restrictions on size, or special fees and tariffs.

Standard & Fair Interconnection Standards to the Grid Interconnection standards set the rules and fees for connecting a customer generator to the grid. The standard should encourage the development of customer systems, while maintaining the safety and integrity of the grid. A fair and reasonable standard has been broadly vetted and adopted in the leading renewable energy states and should be adopted nation-wide.

Solar Fair & Friendly Rates & Utility Revenue Practices Properly designed rates can support investment in solar energy and wise use while maintaining utility profits.

Critical Factor #2—Technology Improvements, including Improvements in Business Model

Technical Cost and efficiency, especially of components have been perennial obstacles to widespread use of solar energy. Both of those concerns have been and are being addressed with incremental improvements. In addition, major improvements are possible in the in the six- to 15-year time frame as research and development initiatives currently in process begin focused commercialization.

Two areas of consideration that have not received as much attention in the past are storage strategies and intelligent control technologies that facilitate integration of renewable energy into the grid. Storage is important for solar energy. It expands its flexibility by extending access to the power produced during sunlight hours. Storage schemes can be grand and large, like compressed-air energy storage. Because of scale and site limitation this approach is not being actively integrated into deployment projects. Other forms of storage such as flow batteries, inverter based micro storage, and flywheels are being considered. The storage industry is currently at a stage of development very similar to where PV was less than 10 years ago. Low volume market demand has meant low volume manufacturing and all of the cost premiums that entails. Properly incentivized storage options will bring investment and scale to its manufacture with the concomitant cost reduction. From the perspective of solar energy deployment in the near-term, these forms of storage should receive both research and incentive support, while core research continues on large scale, big bite strategies that will not be functionally available for five to ten years. Intelligent controls are especially useful to maximize the potential benefits of DG deployment to the grid. Such controls could allocate generation resources across a distribution node to maximize value. Many of these controls are inverter based and would require software and minor hardware additions and modifications, a low cost solution with significant benefits. In addition to technical changes (including modifications of UL 1741), these benefits could be best achieved through a development mechanism of aggregating individual DG sites. There are some fundamental business model improvements that need to be made to integrate DG into the existing utility business model in a way that protects revenues and existing asset base.

Technical Improvements in Business Model One of the most important innovations supporting rapid deployment of solar energy has been the technical advance in business models. The solar industry's explosive growth in the last few years has been directly related to the development and use of the solar power purchase agreement (PPA). In 2007 over 50 percent of the national nonresidential market for solar electric power was developed under PPAs, up from 10 percent in 2006 (''Solar Power Services: How PPAs are changing the PV Value Chain,'' Greentech Media, February, 2008). The solar PPA essentially finances the up-front capital cost and offers customers the output from PV systems at or below the cost of fossil fuel generation. The solar PPA developer monetizes the federal and local tax credits, facilitates utility incentives and renewable energy credit sales, and designs and implements all business processes to minimize and absorb the risk that the customer would otherwise be forced to assume. These risks include: financial, technology, system performance, construction and regulatory. With discipline and innovation, PPA developers have improved and enhanced the solar photovoltaic transaction across the entire value chain, bringing greater profitability and lower prices to the market place.

It was this customer-centric focus, at a time when customers were reeling from rate increases and pricing volatility that resulted in such an expansion of system installations. The solar PPA using PV technology, offers two financial risk reduction strategies for customers: capital acquisition and future price protection. Under the solar PPA, the developer monetizes all of the incentives and tax credits and through aggregation, secures private sector project financing. Because of the nature of PV technology, especially minimal operations and maintenance requirements once installed, and long-term predictable performance output (PV panels have warranties of 25 years), PV can offer firm prices under contract for 20 years. This means an effective 20-year hedge against rising fossil fuel prices for the customer. It is this hedge against rising electric power prices fueled by resources with uncertain and volatile pricing that has made the PV PPAs so successful.

The next generation of solar PPAs, currently entering the market continues this customer-centric focus, but with the addition of utility-centric features. The recent success of solar thermal technologies in the market place (3,000 MW of solar thermal contracts have been initiated with construction expected to be complete in the next three years) highlights the importance of utility-centric features. Solar thermal is a traditional steam turbine electric power generation process, fueled primarily by solar collectors instead of coal, natural gas or nuclear reaction. The familiarity helps many utility executives more readily consider the solar thermal option. But since the approach incorporates a traditional power block, it shares many of the risks and inefficiencies of indirect, multi-stage conversions of energy: large scale, remote location, transmission dependent, multi-year construction, and big impact financing, performance, and operation risks. For these reasons and more, it is not a technology choice and a development approach that can be relied upon to deliver large volume, rapid deployment of solar energy in the first phase. Large scale, strategic, and multi-year development of solar energy in the distributed generation format, especially in the next 10 year period is essential for achieving the goals of Grand Plan for Solar Energy.

Solar energy is a disruptive technology. Disruptive technologies by definition create risk. But disruptive technologies, like the automobile that replaced the horse and buggy, can offer massive improvements in quality of life and prosperity. What mitigates that risk and transforms it into opportunity is the right technology of doing business, the right business model. Such a model must be both customer-centric and utility-centric. Utility revenue and the remaining life of the massive investment made by investors and rate-payers in conventional generation, power distribution and transmission assets must be protected and maximized as best possible while incorporating solar technologies. But not at the expense of future competitiveness and resiliency.

A major obstacle to massive solar energy deployment, in addition to cost and efficiency, has been conflict. Innovations in the solar PPA, coupled with other innovations in power financing entering the marketplace, are designed to end the current conflict between distributed generation and utility revenue protection while establishing more effective and fair financing for rate-payers. It is in our interest to end the conflict. Much can be gained from strategic deployment of DG: improved system reliability, reduction or elimination of transmission & distribution expenditures, reduction of local congestion, voltage support, low cost to no cost for non-participants, and reduced subsidies. Deploying and integrating generators, smart meters and intelligent controls, energy efficiency, virtual net metering, green tariffs and effective storage will permit greater control of load and generation. It is important to note how great the need is for strengthening and hardening our power grid.

> ''. . .the United States has three times as many power outages of the United Kingdom and over 30 times as many power outages of Japan. Both Japan and the United Kingdom have achieved this reliability in part by investing in 21st century distributed generation technologies-distributed solar, combined heat and power, fuel cells, energy efficiency measures, and other customer-centric market solutions. (as quoted in ''The Materiality of Distributed Solar, ''Jigar Shah, Apt, Jay & Lave, Lester & Morgan, M. Granger. (2006). Power Play: A More Reliable U.S. Electric System. *Issues in Science & Technology. http://findarticles.com/p/articles/mi_qa3622/is_200607/ai_n17174065*)

Finally, just as ''simple, easy'' is a useful guiding principle for technologies and technology systems, it is a good design principle for business models as well. ''Solar PPA 2.0'' can reduce and eliminate the need for complex and copious regulations, mandates and other policy requirements.

Critical Factor #3—Effectively Structuring the Multi-Billion Dollar Investment that will be made by Rate-Payers, Investors and the Government

The solar energy investment envisioned by the Grand Solar Plan is significant and such an investment should be fair to all investors and maximize both direct and indirect benefits. Because solar energy financing consists of several mechanisms: tax advantages, utility sector incentives payments, and private capital different investor groups are coordinated in the transaction. Fairness would recommend that all investors receive benefits that justify the investment made. The federal investment for solar is no different than investments made over the last 100 years for general public access to energy and electric power. From 1943 to 1999 $151 billion was spent by the Federal Government for support to nuclear power, $145.4 billion; solar energy, $4.4 billion; and wind, $1.3 billion. (''Federal Energy Subsidies: Not All Technologies are Created Equal,'' Marshall Goldberg, Renewable Energy Policy Project, Research Report July 2000, No. 11). Clearly, for solar energy to become more broadly available, restructuring of the federal energy investment must be made.

The rate-payer contribution is the area of greatest concern for assuring fairness. At the current level of financing for a residential rate-payer, usually less than $50 per year, when some rate-payers benefit more than others, although contributing benefits for all, there is less need for concern. But for the kind of investment that the Grand Solar Plan would entail, spreading the benefits across all rate-payer classes and all communities is crucial. The DG development approach can distribute the benefits across a broader range of rate-payers and communities.

But the fairness issue still remains. Not all rate-payers are in a position to invest in solar energy systems even with the tax credits and utility incentives available. The new funding mechanisms must be designed so that those who can directly benefit, contribute greater investment. Recent developments in PPAs for the DG market can increase the fairness, if the directive to support utility scale coordinated DG development is made.

2. The Current Marketplace for Solar Energy and Expected Evolution

The current solar energy market has been dominated by DG deployment of PV, although 3,000 MW of solar thermal projects are expected to be built in the next few years. PV deployment over the next five years is conservatively projected to increase 35 percent annually. Worldwide 2007 deployment was just under 3 GW and is expected to increase to 11 GW by 2012. US deployments of approximately 300 MW (SEIA as reported by the *Wall Street Journal* 1/18/08) are conservatively expected to grow in excess of 35 percent annually (internal proprietary analysis). The U.S. market is commonly considered to be the next high growth solar market, anticipating greater consolidation of political will and the necessary regulatory framework at the U.S. federal level.

PV is on track for delivering promised cost reductions. Incremental improvements in silicon pricing, silicon utilization and overall system costs are expected to decrease annually at a consistent, but modest level. This is independent of any major game changing technology or manufacturing process coming on-line. There are cost, performance and manufacturing processing improvements in the pipeline, but it is uncertain when, and at what scale they will enter the marketplace. Commercialization is a highly uncertain process and although it is clear that more attention and investment has been directed toward PV improvements across the entire value chain, it is unclear how soon those improvements will be translated into value.

The Grand Plan calls out thin film and expected cost reductions. In general, I am in agreement, though my colleagues at University of Arizona have pointed out some of the fundamental resource issues from both a supply and a toxicity perspective, with the cadmium telluride cells. They and others are working on next generation materials with great promise that avoid supply and toxicity concerns, but again, there are uncertainties concerning time to market.

The greatest concern in the next three to five years may be financing. Solar energy financing comes from multiple sources (federal, utility rate payers and private capital). Difficulties in any of the sectors will constrain the total financing. In particular, without new approaches to utility contributions, in the current near-recessionary (recessionary) environment, there will be limits to how much of a cost burden can be placed on the rate-payer.

PV and HCPV technologies should and will dominate development in the next ten years, especially in a DG format. With all the talk of large-scale projects and exporting to the rest of the country, it would make sense to take care of the domestic needs of potential power exporters first and then use the fixed cost clean power to build generation for export.

The Grand Solar Plan suggests equal development on a GW basis for each technology for the next ten years. On that basis alone, with 3,000 MW of solar thermal in process, the focus for the next five to 10 years should be on PV deployment, especially in the DG development model (PV because it is scalable, modular and flexible can be developed in a central station format or distributed format).

Another way to think about it is to emphasize the technology that offers the "twofers" or perhaps more elegantly "positive externalities." These are other positive benefits that come from the technologies, independent of clean, cost-effective energy generation. Economic development and job creation is essentially the same for each technology, more maintenance jobs for solar thermal, more flexible job experience for PV. Solar thermal is not less expensive than PV, and there is evidence that it is more expensive when comparing scale to scale. Other features and comparisons will be discussed later.

PV is scalable and flexible and it can be developed across a whole range of sizes from a few kilowatts to 50+MW on rooftops and ground mounted, using land that may not have any other productive use. Developed near to the customer demand, transmission costs can be saved. Larger numbers but smaller installations spread across more communities could be deployed, permitting more people and more communities to participate in the economic benefits of a large infrastructure development campaign. Large scale DG deployment also offers additional reliability as has been noted above.

Scale development of any solar technology has the potential to bring cost down, from component manufacturing to installation practices to financing and other transaction costs. Utility scale solar thermal provides component scale benefit solely to utility scale solar thermal. Because the same components are used in PV small scale to utility scale, any wins in the PV area have benefits across the whole range from utility scale down to the small systems on homes or for remote emergency applications.

For the next three to five years it is critical that we allocate solar energy investment to the highest benefit lowest risk installations. That would suggest a predominant role for DG, where the rate-payer investment can be more effectively stretched with private capital, and where the investment has the biggest return to rate-payers: near the load, dispersed throughout communities, benefiting more communities. Larger investments, with longer construction periods, greater cost of construction exposure, higher technology and performance risks, are less beneficial under the current constrained conditions. Since PV can go to scale in a DG format (5 to 20MWs) and at the higher MW level, deliver price breaks equal to or below the current cost of large scale solar power, it is prudent to focus on PV technology and DG scale.

As presented in the technology obstacles above, DG presents difficulties for utilities concerning revenue loss. With the entry into the market place of means to address those concerns, large volume strategically developed and integrated DG projects utilizing PV technologies will dominate in the next 5 years. In the second five, CPV and HCPV advances in scalability will support additional DG and more cost effective central station from regions like Arizona that have good solar resource and available land.

3. Current Regulatory Environment and Incentive Structure & Large-Scale Solar Development

Key to large scale development in the near-term is the extension of the Federal Investment Tax Credit, standard interconnection and net metering at the federal level, support for solar energy on federal lands and protection for the key solar energy financing mechanism to date, the Power Purchase Agreement ("PPA").

The rationale for the first two issues have been offered and discussed above. Support for solar energy development on federal lands could be in terms of multipliers for requirements for federal agencies to deploy solar energy on-site and other federal lands as was done in EPACT 2005 (energy production is doubled for accounting purposes), and in reducing the administrative burden for long-term leases, etc. The fourth issue, protection for the PPA, like standard interconnection and net metering has been addressed in many states, but not all. This requirement concerns the ability of PPAs to be offered by solar energy developers without the burden of excessive and unnecessary regulatory requirements and approval. Federally preempting state attempts to prohibit or restrict on-site generation could consist of the following:

> "Provision of electricity from equipment which uses solar energy to generate electricity shall not be considered a sale of electricity for the purposes of any federal, State, or local regulation governing sales of electricity or regulating utility service, provided the sale is to serve load on the premises where the system is located, or on contiguous property."

4. Distributed Photovoltaics, Concentrating Photovoltaics, Solar Thermal Technology Comparisons, R&D Funding & Congressional Actions

Solar energy consists of two kinds of approaches: capturing the sun's photons (solar electric, photovoltaics, "PV") and capturing the sun's heat (solar thermal). These approaches can be developed in two formats: central station and distributed generation ("DG"). Central Station consists of large scale (20MW to GW, multiple square miles), remotely located, and connected to the grid via transmission lines and infrastructure for distances up to hundreds of miles. Distributed generation ("DG") consists of micro generators of hundreds of watts up to 20MW and can be located near the consumer demand. DG does not require transmission infrastructure, and is delivered to the end-user directly through the service panel or in larger systems of multiple megawatts by means of distribution lines and equipment.

Capturing the sun's heat requires components and equipment that is different depending on whether the developed in central station or DG format. Capturing the sun's photons, depends on similar components regardless of small scale or largescale development. (This is particularly true for PV. For concentrating and high concentrating PV, smaller scale may not be effective).

Utility scale solar thermal approaches include parabolic troughs, power towers, and other systems. Most systems concentrate the sun's heat and focus that heat on production of steam to turn electric generators that then produce electricity.

PV, concentrating PV ("CPV" to 100 equivalent suns) and high concentrating PV ("HCPV" in excess of 100 suns) all use semi-conductor material that when exposed to the photons of the sun, directly produce electric current. The concentrating technologies, by means of special lens, dishes and reflective surfaces, effectively multiply the potential electric current from the photon energy of the sun (some proposed CPV systems are hybrids and use heat for energy production, but they are exceptions). Such systems require tracking and sophisticated thermal management. The complexity is offset by the potential to substantially increase the 10 percent efficiency of PV to 20 to 40 percent. As sophisticated tracking and thermal management technologies from other industries, especially the defense industry, migrate to the CPV and HCPV arena, these complexities could be profitably managed. As experience increases the certainty regarding performance, CPV and HCPV can become more viable, especially those that lend themselves to a scalable, modular and flexible development profile. (*Please see pictures of systems following the text*).

A key consideration for assessing the functionality and finance-ability of a technology, is how quickly and efficiently it can be deployed. Finance-ability requires long-term dependable production, either low cost or reasonably predictable operation and maintenance costs, and other minimized risk factors. The following table summarizes risk factors for PV and solar thermal. Due to the limited deployment of CPV and HCPV technologies, the risk analysis was not meaningful.

Risk Factor Reduction	**PV**	**Solar Thermal**
Fast, simple installation	Yes	No
Modular, scalable, incremental installation	Yes	No
Flexible installation (central station, DG, combination)	Yes	No
Low Operations & Maintenance cost	Yes	No
Faster, less production impacts from O&M	Yes	No
Greater cost reduction 2009 to 2015	Yes	No
Storage can be added at any time	Yes	No
Not dependent on fossil fuel support	Yes	No
Little water needed	Yes	No
Economic Development	Yes	Yes
Dispatch-ability (with storage addition)	Yes	Yes
Cost effective at smaller project size & lower capital requirements	Yes	No

There is a feature of solar thermal that may make it more advantageous and that is storage. Adequate storage increases the dispatch-ability and value of solar energy generation. Large-scale storage for solar thermal, supported by fossil fuel generation, is purported to be farther along in the development and reliability cycle than large-scale storage options for PV. Several proposed projects with storage features are expected to be completed in the next three years and will clarify.

Development Format: Central Station and Distributed Generation

PV (and CPV and HCPV) can be developed in a DG or central station format, though nearly all developments to date have been in a DG format. Utility scale solar thermal requires a central station format. As has been discussed, there are many advantages to DG, as summarized below:

Benefits Central Station & Distributed Generation	DG	Central Station
Direct customer hedge value	Yes	No
Direct customer access to benefits, more customers benefit	Yes	No
More communities have access to economic development	Yes	No
Strengthens Grid Reliability	Yes	No
Increases Energy Security	Yes	No
Allocates Costs more directly	Yes	No
Reduces transmission costs	Yes	No
Reduces need for transmission investment	Yes	No
Relieves Distribution Congestion	Yes	No
Reduces need for distribution investment	Yes	No
Multiple financing opportunities (asset based, tariff based etc.)	Yes	No

Other Regulatory & Incentive Mechanisms

Pricing Carbon Emissions

Currently, pricing carbon emissions has been done indirectly, through an assumed green value attributed to generation from renewable sources. How these attributes are valued and bought and sold is dependent upon the regulatory framework adopted by the state where the project is located. Establishing market based pricing mechanisms at the national level, by means of carbon taxes and/or carbon trading would be very productive and supportive of rapid and efficient deployment of solar energy. Among other positive results would be a reduction in transaction costs.

Setting Standards & Mandates

Although market driven strategies are always to be preferred on core resource issues, standards and mandates are often prudent and necessary to achieve certain objectives. The electric power industry is a regulated monopoly and does not operate in an environment where competitive alternatives can be easily presented and adopted. This is especially true in a market where many of the negative costs have not been systematically included, as is true for electric power. A national requirement or standard for renewable energy deployment could be helpful.

Summary of Federal Research & Development Support and Regulations

Research & Development

1. Storage *Large scale and small scale: batteries, inverter based, flywheels, compressed air storage.*
2. Intelligent Controls for Grid Integration
3. Value and Integration of Distributed Generation
4. Photovoltaic Materials, *including CPV and HCPV*

Regulations

1. Extension of Federal Investment Tax Credit
2. Federalizing Standard Interconnection, Net Metering, and PPA protection
3. Access to federal lands for solar energy deployment
4. Pricing Carbon Emissions
5. Setting Renewable Energy Requirements

In conclusion, we are walking a tightrope of opportunity in the decisions we will make on cleaning and greening our electric power system. And the consequences of making a large, monolithic bad choice are no longer minor. At the end of the day, it all comes down to limiting our risk. Our choices must reflect a hard-nosed look at the risk, no matter how brutal the facts are.

Thank you Madame Chairman and the Members of the Committee for the opportunity to share these observations and opinions with you.

Thanks to my colleagues at Arizona Research Institute for Solar Energy, AzRISE, University of Arizona (Joseph Simmons, Ph.D., Ardeth Barnhardt); SunEdison (Jigar Shah, Greg Ashley, Colin Murchie, and Howard Green), and Raytheon Missile Systems (John Waszczak, Ph.D., and Thomas Olden) for their insights in preparing this document.

Venture Catalyst Inc. is solely responsible for the opinions expressed here.

Statement of Ms. Barbara D. Lockwood, Manager, Renewable Energy, Arizona Public Service Company

Ms. Lockwood. Madam Chairman, Members of the Committee, and staff, thank you for the opportunity to provide APS' perspective on utility-scale solar power.

Arizona is the second fastest growing state in the country, growing at three times the national average. APS serves more than a million customers, who at their peak consume more than 7,000 megawatts of electricity, and electricity demand is growing at a rate of hundreds of megawatts each and every year.

As has been discussed all afternoon this afternoon, in Arizona our most abundant resource is sunshine, and APS is looking for ways to put the sun to work providing electricity. APS is committed to making Arizona the solar capitol of the world.

The focus of my comments today is on CSP or Concentrating Solar Power technology, which we have also heard a lot about from the previous witnesses today.

As Congressman Mitchell mentioned, APS recently announced the Solana Generating Station. Solana is a 280 megawatt solar power plant, to be located just outside of Phoenix, Arizona. APS has signed a long-term contract with Abengoa Solar, the project developer and owner, for all of the electricity generated by Solana. If operating today, Solana would be the largest power plant, solar power plant, in the world.

The plant will use nearly three square miles of parabolic trough mirrors and receiver pipes, and operating at full capacity the plant will produce enough electricity to power 70,000 homes.

Also mentioned frequently today, including Congressman Hall and Chairman Gordon, is the importance of energy storage when it comes to solar, and that is one of the most important aspects of the Solana Generating Station, is its ability to capture and store energy for later use. By using large insulated tanks filled with molten salt, heat captured during the day can be stored and used to produce electricity when the sun is no longer shining. This value cannot be underestimated. Because it can provide energy even after the sun has set, this technology provides maximum value and reliability for APS and its customers.

Solana also provides significant economic benefits for the State of Arizona. The Solana Generating Station will provide 1,500 construction jobs, and 85 permanent operations jobs. The total economic impact is much greater. All totaled, Solana will result in over a billion dollars in economic development for Arizona.

And, Solana is not the end of APS' interest in CSP, as Kate mentioned earlier we believe this is a viable commercial technology that can provide significant energy for our customers in coming years. Depending on many factors, APS alone could envision over 1,000 megawatts of CSP in our system in the next 10 to 15 years.

Today, the single biggest obstacle in the success of Solana is the potential expiration of the 30 percent investment tax credit. I am sure that is something you have heard many times before, and will probably hear many times again. Without this tax credit, Solana is simply not affordable today.

I also need to be clear that a one- or two-year extension of the ITC will not be sufficient. While it may be acceptable for smallscale solar projects, and for wind projects, large-scale solar is different. The approval, permitting and construction of the Solana Generating Station will take three to four years to complete. We cannot begin until we know it will be eligible ITC once it is complete.

As Congressman Mitchell mentioned, if a long-term extension of the ITC is not granted, Solana will not be completed. If the ITC is extended for a sufficient period, there will be many other plants like Solana built in the next five to 10 years. If not, the industry will lose its momentum and no large-scale solar plants will be constructed.

The future of large-scale solar depends on getting those first few plants in operation.

Thank you, Madam Chairman, and Members of the Committee, for the opportunity to share this information with you.

[The prepared statement of Ms. Lockwood follows:]

Prepared Statement of Barbara D. Lockwood

Mr. Chairman and Members of the Subcommittee, thank you for the opportunity to provide Arizona Public Service Company's (APS') perspective on utility-scale solar power. My comments will focus on the opportunity solar provides for clean, reliable electricity, and the challenges associated with realizing that potential.

APS is the largest and longest serving electric power utility in Arizona. Arizona is the second fastest growing state in the country, and APS has more than a million customers who, at their peak energy consumption, use more than 7,000 megawatts of electricity. By 2025, APS will have nearly two million customers demanding over 13,000 megawatts of electricity. To meet this rapid growth in electricity demand, APS is investing $1 billion a year in infrastructure. That number does not include additional generation sources. For APS alone, our peak demand is growing at hundreds of megawatts per year, or the equivalent of one medium-sized natural gas plant each and every year. Meeting the growing needs of our customers is both a challenge and an opportunity.

In Arizona, our most abundant renewable resource is sunshine. The solar resource in Arizona is virtually unlimited, with more than 300 days of sunshine each year. In addition, Arizona has sizable quantities of wide-open, flat landscape that is ideal for the installation of large-scale solar equipment. Among the most important factors in considering a resource for electricity production is the reliability of the fuel. Arizona's solar climate provides a resource that is both dependable and predictable.

APS is committed to making Arizona the solar capital of the world and bringing affordable renewable energy to all its customers. A balanced renewable energy portfolio including solar, wind, geothermal and biomass/biogas resources is fundamental to our operating strategy. For the past two decades, APS has worked with the solar industry and researchers around the U.S. and the world to bring lower cost and reliable solar electricity to our customers. In 1988, the APS Solar Technology And Research (STAR) center was developed to support the advancement of solar resources, including field operation of both photovoltaic and concentrating solar technologies. In addition to STAR, APS currently has over five megawatts of photovoltaic power plants in operation providing reliable solar energy to our customers.

APS has also supported the advancement of concentrating solar power (CSP). These technologies are ''thermal electric systems'' that use solar heat to drive generators and engines. CSP thermal systems include solar trough concentrator systems and central receiver (power tower) systems that use many mirrors to focus light on a central solar collector. CSP also include solar dish Stirling systems and other advanced solar concepts.

In fact, APS constructed the first commercial CSP plant in the United States in almost 20years. The Saguaro Solar Power Plant, which came on-line in 2006, is a one megawatt parabolic trough facility located just north of Tucson at Red Rock, Arizona. This plant has provided critical learning for APS, the CSP industry, and researchers. While small in size, it has facilitated new interest in CSP around the country and the world.

But that was just the beginning of our entrance into commercial CSP. Also in 2006, APS stepped forward to lead a coalition of southwestern utilities interested in CSP. The Joint Development Group is a consortium of seven entities exploring the possibility of a 250 megawatt CSP project to be located in Arizona or Nevada. Acting as project coordinator, APS issued a request for proposals in December of 2007. If all goes well, the consortium project could be selected this summer.

But our most significant step to date is the announcement on February 21, 2008, of the Solana Generating Station. Solana is a 280 megawatt solar power plant to be located 70 miles southwest of Phoenix near Gila Bend, Arizona. APS has signed a long-term contract with Abengoa Solar, project developer and owner, for 100 percent of the electricity generated by Solana. Solana is the Spanish word for ''sunny place.''

If operating today, Solana would be the largest solar power plant in the world. The plant will use nearly three square miles of parabolic trough mirrors and receiver pipes, coupled with two 140-megawatt steam generators. Operating at full capacity, the plant will produce enough electricity to power 70,000 Arizona homes.

Solana also provides significant economic benefits to the State of Arizona. The Solana Generating Station will provide 1,500 construction jobs between 2008 and 2011 and 85 permanent operations jobs. Solana will also generate between $300 million and $400 million in tax revenue over the 30 year life of the plant. All total, Solana will result in over $1 billion in economic development for the Arizona economy.

Finally, Solana is an emission-free source of electricity, avoiding nearly 500,000 tons of carbon dioxide, 1,065 tons of nitrogen oxides, and 520 tons of sulfur dioxide each year. It is the equivalent of removing 80,000 cars from the road each year. Solana will also use 75 percent less water than the current agricultural usage of the land.

APS selected Abengoa Solar as its partner for Solana because of its track record as a solar developer, its critical operational experience and a reputation for meeting contractual obligations.

One of the most important aspects of Solana is its ability to capture and store solar energy for later use. By incorporating large insulated tanks filled with molten salt, heat captured during the day can be stored and used to produce electricity when the sun is no longer shining. The molten salt and heavily insulated tanks are able to retain heat with very high efficiency, and the stored heat can then be extracted in the evening or even the following day to create electricity.

The stored heat not only increases the total amount of electricity generated, it also adds specific operating benefits for APS. The ability to use stored heat on demand, also referred to as ''dispatching,'' allows APS to respond to customer usage patterns and emergency energy needs more effectively. Most southwest utilities experience their highest customer demand during the summer months. While the power need is substantial in the middle of the day, peak energy demand occurs in the late afternoon and into the early evening hours. Because it can provide energy even after the sun has set, the solar trough with thermal energy storage provides the maximum value for APS and its customers.

Diversification of generation resources is critical to maintaining a reliable electric system and concentrating solar power provides a significant opportunity to diversify energy resources. In addition, the costs to construct and maintain concentrating solar power plants have declined while at the same time equipment and labor costs, rising fuel prices and emissions concerns are increasing the risks of conventional resources.

APS also recognizes that renewable energy strategies will become even more important under the prospects of carbon legislation. With zero carbon emissions, energy from solar power provides one method of addressing concerns around global warming while continuing to provide reliable electricity to our customers.

And Solana is not the end of our interest in CSP. APS is currently engaged in a formal dialogue with our regulators, stakeholders and customers about our future energy sources. We are exploring the availability, cost, regulatory and policy implications associated with many different types of resources including nuclear, natural gas, coal, energy efficiency and renewable energy. One of several scenarios under discussion is one where CSP plays a central role, adding 1,350 megawatts by 2020. Each of these efforts will help us to meet, and possibly exceed, the progressive Renewable Energy Standard established by the Arizona Corporation Commission.

CSP, in particular the solar trough, is proven, reliable technology. There are no technical barriers to deployment of this technology today, and APS is aggressively exploring the near-term potential.

In considering the long-term potential for utility scale solar, one topic of consideration is how to integrate large solar plants into the regional and national electric grid. This topic raises numerous issues including availability of land for large scale installation and the availability of transmission facilities and transmission capacity to deliver the energy to load centers. The lack of transmission capacity and how that is managed will be a significant factor in the long-term success of utility-scale solar. In fact, transmission is generally constrained in much of the west and significant new transmission investment is needed in the coming years for all types of generation be they renewable or conventional generation. New transmission is being planned throughout the west and in California, New Mexico, Nevada, and Texas specifically to access renewable resources including wind and geothermal. Others states and utilities, including APS, are studying their needs for both intra- and interstate transmission to ensure a robust grid to meet the needs of the West's burgeoning population. The studies include the ability to reach those areas of the west with abundant cost-effective renewable resources.

Also, the possibility of locating large scale solar on federal land should be investigated and analyzed. By its nature, solar technologies require significant geographic footprints. A general rule of thumb for a solar installation is five to 10 acres per megawatt. As I previously stated, the Solana Generating Station requires three square miles of contiguous land. Considering that the Federal Government is the largest land owner in the U.S., a study of federal land in high solar resource areas that may be made available for CSP development would also be beneficial and appropriate.

However, the biggest obstacle to the success of utility-scale solar, including Solana, is the potential expiration of the federal Investment Tax Credit (ITC). Solana, and projects like Solana, became possible when the federal ITC for solar systems was increased from 10 percent to 30 percent in 2006. While large-scale solar is still more expensive than conventional resources, the 30 percent investment tax credit decreased the cost sufficient to make these projects a reasonable option. Without these tax credits, large scale solar projects, including Solana, are simply not affordable today. As you know, the 30 percent ITC is scheduled to expire at the end of 2008. The approval, permitting and construction of the Solana Generating Station will take three to four years to complete. The Solana project also requires well over a billion in capital investment. APS, Abengoa Solar, and the financial institutions providing funding for Solana require certainty that Solana will be eligible for the ITC once operational. *If a long-term extension of the ITC is not granted, Solana will not be completed.*

A different federal tax credit, the production tax credit (PTC), has spurred significant development for other renewable energy resources, most notably wind energy. The PTC has been extended five times since its introduction in 1992 and each extension was for one to two years. Although the wind industry has worked toward longer-term extensions, wind energy projects, and smaller scale solar projects, have much shorter time frames for construction, which makes short-term extensions of the PTC acceptable, if not preferable. Although the solar ITC is typically packaged with the PTC in discussions of extensions, large-scale solar has very different needs related to tax credits. *A one- or two-year extension of the solar ITC is simply not sufficient to make large scale solar projects like Solana a reality.* In fact, a one or two year extension of the Solar ITC may effectively cancel the project. Large scale solar has little hope of realizing its potential without a long-term extension of the ITC. APS believes an eight-year extension is optimal. Eight years should be sufficient to get a number of large scale solar facilities completed. It is also long enough to establish the supporting industries like mirror and receiver manufacturing in the United States. Once the industry gains a foothold, prices will decline and incentives will no longer be necessary.

Another critical aspect of the ITC is the fact that it is not available to public utilities. The restriction needlessly narrows application of the credit and is unfair to U.S. citizens because the vast majority purchase power from a public utility, as it is defined by the tax code. This current policy forces a third-party owner to take advantage of the ITC and it creates unnecessary uncertainty and costs to the system. It requires the utility and regional grid to consider the operational and financial risks inherent in any third party relationship thus potentially affecting the utility operating strategies. APS is managing these risks with Solana, but it creates a sub-optimum situation when it is the only strategy available.

I was also requested to address a recently published article. ''A Solar Grand Plan,'' published in the *Scientific American Magazine* in December 2007, describes a world where solar energy provides 69 percent of the U.S.'s electricity by 2050. It includes huge tracts of land covered in solar and a new direct-current transmission system across the U.S. It also includes 16-hour thermal storage for CSP and compressedair energy storage for photovoltaics, which allow the production of energy from solar resources around the clock.

''A Solar Grand Plan'' is certainly grand. It's a big, bold vision for a new energy era. Without analyzing the details of the plan, there appear to be no glaring technical issues with the proposed strategies. CSP and photovoltaics are proven technologies. As described, thermal storage and compressed-air energy storage are likely viable concepts. Finally, direct current transmission is already in operation today.

No, the challenges with this plan are not technical. But there are enormous planning, regulatory, and policy challenges with achieving this vision. Most importantly, energy policy decisions are made largely at the individual utility and State level. Each utility and state has different perspectives, and different regulatory authorities, that control the vast majority of decisions around generation sources and transmission. And although I haven't analyzed the cost presented in the article, the execution of such a plan would clearly depend on gaining great cost efficiencies.

Clearly, the potential for utility scale solar electricity is enormous. If, and only if, the ITC is given a long-term extension, I predict we will see several thousand megawatts of utility scale solar developed in the next five to 10 years. At least seven major projects have been announced since 2006. If the ITC is not extended for a sufficiently long period of time, the industry will lose its precious momentum and no large scale solar plants are likely to be constructed. The future of large scale solar depends heavily extending the ITC and getting those first few plants in operation.

These initial plants are planned to supplement existing fossil fuel resources and help to satisfy our growing energy needs. In the long-term, utility scale solar could be a viable option in replacing base load fossil fuel facilities as those assets are retired. But costs need to decline significantly to make that a viable option. Only then will solar be a viable option for replacing base load assets that are being retired. Assuming success in the near-term, the prospect for the next 20 to 50 years is virtually unlimited.

Thank you, Mr. Chairman and Members of the Subcommittee for the opportunity to share these observations and opinions with you.

BIOGRAPHY FOR BARBARA D. LOCKWOOD

Barbara D. Lockwood is the Manager of Renewable Energy for Arizona Public Service (APS) where she is responsible for APS' renewable energy programs including large-scale generation and customer programs. Ms. Lockwood joined APS in 1999. Ms. Lockwood began her career in the chemical industry at E.I. DuPont de Nemours in various engineering and management roles on the east coast. Subsequent to DuPont, Ms. Lockwood moved into consulting and managed diverse projects for national clients across the United States.

Ms. Lockwood holds a Bachelor of Science in Chemical Engineering from Clemson University and a Master of Science degree in Environmental Engineering from Georgia Institute of Technology. Ms. Lockwood is a registered professional chemical engineer in Arizona and California.

Ms. Giffords. Thank you, Ms. Lockwood.

And finally, Mr. Kastner.

STATEMENT OF MR. JOSEPH KASTNER, VICE PRESIDENT OF IMPLEMENTATION AND OPERATIONS, MMA RENEWABLE VENTURES LLC

Mr. Kastner. Thank you, Madam Chairman, and the Subcommittee Members, for providing this opportunity.

As many of your questions addressed the Nellis Air Force Base project I would like to begin by recognizing Lieutenant Colonel Karen White, who is in the audience today. Lieutenant Colonel White was instrumental in making the Nellis Air Force Base project a success, and it is great to have her here.

Ms. Giffords. Welcome, Colonel, thank you for being here today. We appreciate it.

Mr. Kastner. On behalf of MMA Renewable Ventures and the solar industry, I am happy to provide the following comments related to the development and financing of utility-scale solar.

In 2007, MMA completed the project development and third party financing of over 22 megawatts of solar, including the 14 megawatt facility at Nellis Air Force Base, the largest PV installation in North America.

We are actually pursuing domestic opportunities in other renewables and our parent company MMA has built this business largely around sustainable and socially-responsible investment opportunities, including affordable housing, renewable energy, and sustainable land investments.

Ms. Giffords. Mr. Kastner, I don't mean to interrupt, but can you move the microphone a little bit closer?

Mr. Kastner. Sure.

Ms. Giffords. Okay.

Mr. Kastner. The Nellis project involves a public/private partnership that is advantageous because the affiliated public entities could not avail themselves of the Federal tax benefits. The project was enabled by the following commercial arrangements:

1. A 20-year site lease with the United States Air Force;
2. A power purchase agreement with Nellis Air Force Base;
3. An agreement to sell Nevada Power the renewable energy credits associated with the project for 20 years;
4. An installation contract with Sun Power Corporation;
5. And finally, financial arrangements that included construction financing from Merrill Lynch, permanent debt financing from John Hancock, and tax equity financing from Citicorp, Allstate and MMA Financial.

The combination of these complex legal and financial arrangements enabled the project. We believe that this type of public/private partnership provides a commercial approach that can be used at a variety of sites of varying size and scale in the U.S.

All fuel-less electric generation technologies are more capital intensive than conventional combustion-based technologies. And thus, they require long-term stability and certainty in financial, legal and regulatory environments, in order to mobilize the long-term investment. The following concepts are key to providing stability and certainty for solar projects.

As has already been mentioned, first and foremost, a long-term extension is needed for the current 30 percent investment tax credit. The ITC is critical to the development and financing of utilityscale solar projects and DG projects, and necessary to ensure continuing domestic project development.

History has shown that the short-term tax credit, subject to the uncertainty of congressional reauthorization, can actually be detrimental to the development of renewable energy. Uncertainty around the extensions of the production tax credit for wind power increases the cost of capital for projects and causes the inflation of equipment costs due to supply constraints. We would expect the same fate for solar without the long-term extension of the tax incentive.

Stability and certainty are also critical for the commercial arrangements which enable the projects. In the case of the Nellis installation, a change in law risk within the standard REC contract had the potential to make the project somewhat less than financeable. If the Public Utility Commission of Nevada had not issued an order that provided assurances regarding this change in law risk the project may not have been financed.

In most states where solar PV projects are being implemented, they are enabled, at least in part, by State renewable portfolio standards. This is certainly true for the Nellis project.

The geographic reach and timing of such programs would be enhanced through a Federal initiative. The National RPS would provide certainty by guaranteeing a minimum degree of market demand for renewable energy. Such an initiative must provide ample flexibility for

State programs that surpass federal minimum standards and encourage the dissemination of best regulatory and utility procurement practices, including standardized contracts that provide sufficient long-term certainty for mobilizing capital markets.

Such an initiative should also support diversification through the development of promising technologies in appropriate regions. A specific solar requirement from the Nevada RPS helped to enable the Nellis project, and is why it occurred in Nevada, and not in Arizona or New Mexico.

Lastly, a federal cap and trade system, or emission tax, would help to internalize the environmental and social costs of emissions caused by burning fossil fuels. This would also help to create a level playing field for all generation types.

In conclusion, investors are beginning to respond to the market-driving incentives for solar energy provided by Federal and State government. The Nellis project is a great example of how these types of incentives can be combined to create a viable project opportunity.

These types of opportunities will only reach the volumes required to significantly reduce the costs of solar energy if the incentive programs are structured to ensure the creation of a stable, long-term market for investment. The market for these opportunities could be expanded greatly through various actions at the federal level, including a national RPS and the adoption of a market mechanism or tax for internalizing external costs of emissions from conventional energy sources.

Thank you.

[The prepared statement of Mr. Kastner follows:]

Prepared Statement of Joseph Kastner

Good Morning. On behalf of my company, MMA Renewable Ventures, LLC and the solar industry, I am happy to provide the following comments related to the development and financing of utility scale solar projects.

In 2007, MMA Renewable Ventures completed the project development and financing of more solar photovoltaic projects in the United States than any other company in the U.S. as measured by total capacity installed (more than 22 MWp) from over 20 discrete projects. We are especially proud of the development and financing of the 14MWp solar photovoltaic (PV) project on Nellis Air Force Base—the largest such project ever built in North America and one of the largest in the world.

In a significant portion of these projects, the land owner and power purchaser has been a public entity such as a Federal Government department, municipality, or transit district. As you know, such entities cannot avail themselves of the federal investment tax credits (ITCs) and accelerated depreciation benefits offered under the Internal Revenue Code. In all of these transactions, MMA Renewable Ventures served as the third party project developer and financing party which develops the projects, negotiates the power purchase agreements, secures the necessary land rights, negotiates engineering, procurement, and construction contracts, negotiates interconnection agreements with distribution utilities, and obtains construction and permanent financing (debt and tax equity). Consequently, we are intimately knowledgeable and experienced with every aspect of project development and finance of solar PV projects.

In addition to solar PV projects, MMA Renewable Ventures is actively pursuing and developing wind, biomass, biofuel, and energy efficiency opportunities in the U.S. market. We expect to add energy efficiency projects to our portfolio of operational assets in 2008 and wind, biomass, and biofuel projects within the next two calendar years. Similar to solar, many of these opportunities are dependent upon an extension of currently existing tax credit provisions, in this case the production tax credit (PTC) in Section 45 of the tax code.

MuniMae, the parent company of MMA Renewable Ventures, has built a business largely around sustainable and socially responsible investment opportunities. Historically, this has involved affordable housing and more recently renewable energy and sustainable land investments.

Description of the Solar Project at Nellis AFB

The development and financing of the solar project at Nellis Air Force Base (AFB) was based on the following commercial arrangements:

1. Nellis AFB has leased 140 acres of property to a special purpose entity called Solar Star NAFB, LLC, owned and operated by MMA Renewable Ventures, for a period of twenty years beginning January 1 following the start of commercial operation for the project;
2. Solar Star NAFB has in turn agreed to sell the power output of the plant to Nellis AFB for a coincident term;
3. Solar Star NAFB has also agreed to sell the renewable energy credits (RECs)—the tradable credits representing the environmental attributes, benefits and other values of renewable energy—to Nevada Power for the same 20-year term. Nevada Power purchases such credits in order to comply with the Renewable Portfolio Standard required under Nevada's Renewable Energy Law;
4. On behalf of Solar Star NAFB, MMA Renewable Ventures negotiated an engineering, procurement, and construction contract (EPC Contract) with PowerLight Corporation, which is now SunPower Corporation, Systems (SunPower). Under the EPC Contract, SunPower purchased more than 70,000 solar modules and 54 inverters, constructed the tracking systems, assembled racks of modules, transported equipment, arranged labor on the site, and interconnected all the system components;
5. On behalf of Solar Star NAFB, MMA Renewable Ventures arranged for construction financing from Merrill Lynch, debt financing from John Hancock Insurance Company, and equity financing from CitiCorp North America, Allstate Insurance Company, and MMA Financial.

The sum total of these complex legal and financial arrangements enabled the construction of the largest PV plant in North America. While the specifics of each party and arrangement may vary from project to project, we believe that this public-private partnership model provides a commercial approach that can be used at a variety of sites of varying size and scale.

Recommendations for Promoting Utility-Scale Solar Projects

Utility-scale solar projects represent the greatest opportunity for solar electric generation technologies to reach cost parity with conventional gas and coal-fueled electric generation.

When equipment, labor, and capital are deployed to build solar projects at a scale counted in tens of megawatts, gains from economies of scale including the spread of transaction costs can deliver lower cost solar power. Additionally, this will spur the cost efficiencies required to make the deployment of distributed generation more competitive with retail electricity rates requiring minimal subsidies.

In order to promote the development of projects of such scale, project developers and financial entities need to have a relatively stable financial, legal, and regulatory environment. All fuel-less electric generation technologies are more capital intensive than conventional combustion-based technologies, requiring long-term stability in the business environment to mobilize capital. The following concepts/initiatives are key to the development and financing of utility-scale solar:

1. Long-Term Federal Tax Incentives

The current 30 percent investment tax credit (ITC) for solar projects expires at the end of 2008. At present, these federal incentives are critical to the development and financing of utility-scale solar projects. Without the federal tax benefits, utilityscale solar projects will not be viable because the cost of energy will simply be too high.

The effectiveness of existing incentives is significantly limited in driving development of utility scale projects with long lead time particularly given the pace of development and consumer adoption of energy technologies. The existing tax credits or incentives are short-term, piecemeal programs subject to the uncertainty of the Congressional reauthorization and/or appropriations processes. For example, the production tax credit for wind and other types of renewable energy, established in 1992, has been subject to three expirations and several short-term extensions (some retroactive). Uncertainty around the ITC extension increases the cost of capital due to the risk of meeting a deadline and leads to a boom and bust cycle which has caused the inflation of equipment costs purely from supply constraints.

Congress should pass a long-term ITC to drive substantial private sector investment in clean energy technologies. Investors need stable, long-term, and predictable incentives. MMA Renewable Ventures supports a minimum seven-year timeframe for clean energy tax credits because this is the minimum period necessary to enable rational investment decisions and deployment of resources in utility scale projects. The federal regulatory environment's support for energy technologies can be significantly improved by establishing consistency and predictability.

At the bottom line, those of us who are actually building and financing utilityscale solar projects need greater certainty of the federal tax benefits. In addition, the ITC could benefit from the amendment of several rules within the IRS code:

- Eliminate the basis adjustment so that one-half of ITC is not "recaptured";
- Make renewable energy investments eligible for Community Reinvestment Act (CRA) consideration. Structured correctly, this could serve to catalyze both distributed and utility-scale solar projects in low and moderate-income communities and/or serving public facilities. It would also serve to attract additional institutional investors into the space and help to create "green-collar" jobs in lower-income communities;
- Create an "economic substance" carve-out for solar tax credits similar to what was done for low-income housing tax credits;

- Raise the production tax credit (PTC) for solar to make it competitive with the ITC and give investors a choice of either one. The PTC structure is a better fit for some investors and will encourage more capital to enter the solar space;
- Match the residual value exemption currently available to the low income housing sector, allowing for no constraints at resale after the tax benefits have been monetized;
- Abolish the possibility for ITC recapture in the event of a catastrophic loss without replacement by the end of a calendar year;
- Allow tax equity to enter project after the system as reached commercial operation under any financing structure.

2. A Stable Legal Framework

One of the important prerequisites for investors in utility-scale solar projects is certainty the commercial arrangements will remain intact for the full term of the financing. Utility purchasers, commissions, and State and federal regulations all need to provide certainty and assurances that the various commercial arrangements will not materially change throughout the life of the project.

For instance, in reviewing the standard contracts proposed for the Nellis AFB project it was determined that certain elements in the site lease and the streams of revenues from the power purchase arrangement with Nellis AFB and the REC Agreement with Nevada Power made the project somewhat less than financeable. The most significant instance involved the change-in-law risk associated with the REC agreement. If the Public Utility Commission of Nevada had not issued an order that approved the contract and an associated stipulation that provided assurances regarding change-in-law risk, the project might not have been financed.

3. A National Renewable Portfolio Standard

Today, renewable energy resources provide a fraction of total U.S. energy, with the potential for significant growth. More than twenty-seven states and the District of Columbia utilize a wide variety of renewable portfolio standard (RPS) mechanisms to drive a greater reliance on renewable energy. A basic RPS requires the electric utilities (investor-owned utilities and publicly-owned utilities) within a state to procure a percentage of their electricity output necessary to meet load from renewable energy sources in a specified timeframe. Current State policies require varying percentages of renewables, typically targeting a goal of one percent to five percent in the first year, increasing each year to achieve a goal of five percent to 20 percent over approximately 10–15 years.

In general, a utility can meet RPS requirements by incorporating renewable energy into its fuel mix in one of four ways: (1) building renewable energy facilities; (2) purchasing power directly from an existing renewable energy source; (3) buying RECs; or (4) by encouraging production of distributed renewable energy, efficiency, or conservation. The specifics of each RPS program vary widely state to state from the goal, to the criteria, to the method of implementation. Many State programs set standards for specific technologies to ensure diversity of electricity supply by supporting the development of promising technologies that may not currently be the most economic.

A national RPS would set the minimum standard for wholesale renewable energy usage throughout the United States. This would serve the important function of guaranteeing a minimum degree of market demand for renewable energy generation. Every state would be required to develop an energy regulatory strategy that includes a base level RPS with performance-based metrics that would drive investment in, and adoption of, viable, cost-effective renewable energy technologies. Specifically, Congress would mandate the establishment of minimum State renewable energy procurement standards with ample flexibility for State programs that surpass the federal minimum standards, encouraging dissemination of best regulatory and utility procurement practices, and providing states with incentives to increase reliance on renewable energy, reward energy efficiency, and to provide for a national REC market. For the reasons stated previously regarding stability, it is important that a national RPS is cognizant of existing State programs to ensure long-term investments already undertaken are not adversely affected. The federal RPS would require sufficient non-compliance measures in order to provide a strong incentive for utility compliance.

A national RPS can be a market driving, demand side solution for addressing the broader goals of energy policy through development of diverse, secure renewable energy sources and energy efficiency, while at the same time encouraging technological advances throughout the energy supply chain. The future of renewable energy production in the United States resides in this synergy of governmental policy and emerging technologies—and without each, the aim of diversified, sustainable, and efficient energy production is simply impossible in the foreseeable future. By setting these aggressive goals for renewable energy production targets, the government will drive innovation and the market will create solutions.

4. Valuing Carbon Emissions and Other Externalities

The current cost of conventional fossil-fuel electricity does not include the environmental and social costs associated with the emission of carbon, mercury, and other pollutants into the atmosphere. Either a cap-and-trade system or emission specific taxes would complement long-term subsidies and the establishment of minimum market demand by internalizing the impact of burning fossil fuels into the price of electricity. This would tend to make solar energy more competitive with fossil fuelfired electricity and further boost investment.

Market Differences in the Southwest

The southwestern portion of the U.S. including California, Nevada, Arizona, and New Mexico has the strongest solar resource in the country. The State of Nevada has an RPS-driven REC market that provides a large part of the economics for the Nellis solar project. The RPS rules for the state have specific requirements for solar and applies a multiplier to RECs (termed Portfolio Energy Credits under the Nevada RPS) produced by solar facilities. RPS programs in the other three states exist, but are not necessarily structured properly for significant market penetration of utility-scale solar projects.

California

California has catalyzed solar development through the California Solar Initiative (CSI) program which utilizes a short-term production based incentive. This direct subsidy has spurred the development of distributed generation projects (mostly less than one megawatt), but is not applicable for utility-scale projects. It is expected that California will introduce a tradable REC program for the existing state RPS in the near future that will encourage distributed generation projects currently suffering from subsidy levels declining faster than capital costs for key equipment. California utilities have utilized a request for offer (RFO) process fulfilling their RPS requirements. Since there is no solar set-aside, most of these contracts have been awarded to other renewable technologies that are currently more cost effective than solar. Contracts which have been awarded to solar projects under the RFO process have largely gone to earlier stage solar technologies that have yet to be implemented. The California Public Utilities Commission recently announced a feedin tariff based on the a revised calculation methodology for the ''market price referent'' that sets the ceiling price for contracts awarded in the RFO process. The new methodology attempts to take into account the time-of-use benefits associated with the solar production curve matching well with the state-wide demand in California. The current consensus is that the announced feed-in tariff does not provide adequate levels of compensation for solar PV projects.

Arizona

There are certain regulatory hurdles that impede the development of solar and other clean technologies in Arizona. Low energy rates and tariff structures that do not adequately incentivize the peak-producing benefit of solar negatively impacts the economics of solar. Net-metering policies are essential to opening up the market to more wide-spread adoption, instead of limiting potential customers only to those who have 365 day operations, and large load centers. Under the current net-metering rules only small systems are rewarded, otherwise solar generation that exceeds on-site usage is not compensated for. Like net-metering, interconnection standards must be standardized across the state and have a minimum of 2MW to sufficiently promote industry adoption. Lastly the available incentives are insufficient. APS has taken the lead in establishing a PBI program, which is an important step, and for the most part well-designed (20 year PBI structure), however the total available funding is only enough to fund a few MW per year—which is not enough to entice the solar PV industry to undertake the cost and risk of entering a new market.

New Mexico

New Mexico has shown true leadership in the aggressive RPS goals and high net metering limits. This includes solar specific requirements that must be fulfilled beginning in 2011. The law also includes a ''Reasonable Cost Threshold'' which limits the payment of power from solar installations to currently unfinanceable levels.

Conclusion

Investors are beginning to respond to the market driving incentives for solar energy provided by Federal and State governments. The Nellis AFB project is a great example of how these types of incentives can be combined to create a viable project opportunity when a third-party can enter and efficiently monetize the tax benefits. These types of projects will only reach the volumes required to significantly reduce the cost of solar energy if the incentive programs are structured to ensure the creation of a stable, long-term market for project developers, installers, equipment manufacturers, and investors. The geographic market for these opportunities could be expended greatly through several actions at the federal level including a national RPS and the adoption of a market mechanism for internalizing the external costs of emissions from conventional sources of energy.

BIOGRAPHY FOR JOSEPH KASTNER

Joseph Kastner is Vice President of Implementation and Operations for MMA Renewable Ventures LLC. He is responsible for sourcing and developing qualified renewable energy projects that fit the investment profile of the company and oversees the management of assets under construction and operation. Prior to joining MMA Renewable Ventures, he was responsible for project implementation, operation and maintenance as the renewable energy division manager for NUON Renewable Ventures USA LLC, a U.S.-based subsidiary of the Dutch utility NUON bv. Prior to NUON, Mr. Kastner worked as a consultant to commercial and residential building owners and investors in the areas of energy efficiency and the use of photovoltaic and thermal solar energy systems. Mr. Kastner has a Master's Degree in Energy Engineering from Stanford University, a Master's Degree in Environmental Science and Management from the Donald Bren School at the University of California–Santa Barbara, and a Bachelor's Degree in Mechanical Engineering from the University of Minnesota.

DISCUSSION

Ms. Giffords. Thank you, Mr. Kastner.

We appreciate the testimony from all of our witnesses, and now we are going to turn the floor over to the Members of the Committee to have a chance to ask you some questions.

I know that Mr. Hall has to leave shortly after 2:00, so I believe if I can get my questions in first, Mr. Hall, then we will move to you, and I know you have to catch your plane.

In terms of the technology on the table, it is set for a briefer period than the five minutes, so what we will do is, halfway through you will see the light turn on, but when you see it getting orange and then red, if you can please either close up the questions, we'd like to be able to move rapidly through a round of questions.

THE GRAND SOLAR PLAN: JOBS AND ECONOMIC BENEFITS

So, I would actually like to kick off with Mr. Hansen, who talked about the Grand Solar Plan. For those individuals who have not had a chance to read the ''Scientific American'' article, we can provide that through the office here, and I will make sure that Members of the Committee have access to it as well, although I believe that most Members have had a chance to read it. It is very compelling. But, there we go, that is the front cover. There we go.

I would like to dig a little bit deeper in terms of the economic benefit. You talked a lot about what it would bring in terms of the energy, and the power, and sustainability, but could you speak a little bit in terms of jobs and actual dollars?

Mr. Hansen. Madam Chairman, I tried to pack about 10 years worth of information into five minutes, so it was difficult.

The Solar Grand Plan, looking from the year 2009 to 2020, is the period when the incentives, the subsidies so to speak, to help move the technology forward, would be needed.

It is our belief that after 2020 that those technologies would be compatible economically with traditional generation methodologies—coal, natural gas, et cetera. So, beyond 2020 the economics would look very similar to how the current economic situation is for development of new electrical generation.

Between now and that point in time, I believe our study indicated there would be about 150,000 jobs that would be generated. And, really, we are looking for the Abengoa type of projects to be part of this Solar Grand Plan. There would be some additional drilling required, some people that would be required to additionally add that capacity for storage in the ground, but that would be, again, existing technology, it would provide more jobs for the natural gas drillers, if you will, to provide that kind of storage in the future.

All of the details have not been shaken out yet, we need more time to be able to put it together. We actually do have another paper coming out. It is out for peer review at the present time, that will have more information of the details of how many jobs and where those jobs would be located. The transmission system, the energy storage systems, they will not be just in the U.S. southwest, they will be throughout the United States. So, it will bring benefits to all of the United States, not just the southwest.

Nellis Air Force Base Partnership

Ms. Giffords. And, going back to Mr. Kastner, and I want to just make it clear for everyone who understands, that Nellis Air Force Base launched in a public/private partnership not more than a couple of years ago, and I believe it took you about a year to complete this project. But, 40 percent of the energy consumed on Nellis Air Force Base is now powered by solar energy, which is extraordinary when you think about a very short period of time and ability to move so quickly in that.

The Mayor and I will be visiting Nellis Air Force Base in a couple of weeks, and we will be bringing members of the community along with us as well. But, can you talk a little bit about, from a policy standpoint, several of us are from the southwest, from southwestern states, you know, what did it really take in terms of leadership to be able to implement that project, tax credits, and how you can see that expanding in different states as well?

Mr. Kastner. Yes. I mean, the key driving force for the economics is the renewable energy credit contract with Nevada Power, and that was really catalyzed by the renewable portfolio standard within Nevada that requires a certain amount of solar to be produced within the state.

Nevada Power issued an RFP for qualifying for these contracts before the Nellis project, during the conception of the Nellis project, and that is what, you know, really brought it forward, provided an RFP by the Air Force to do the project, knowing that this contract was available.

International Competition in Solar Energy

Ms. Giffords. And, a general question I want to address to Ms. Maracas, and to Ms. Rauluk. I see behind you we have some representatives from the Solon AG Company, a German company that has been investing here in southern Arizona, and I want to thank you for that. I know that the United States southwest is now competing, not just with areas like

Nevada, or California, or New Jersey, but we are now competing with different countries as well.

Can you please touch on some of that international competition, and how we here, you know, in the United States are going to be able to be a major player 10, 15, 20, 25 years from now?

Ms. Maracas. Yes. I think that is a very relevant question, and it is worth adding, I think, that one of the reasons that large-scale companies like Abengoa, Bright Source, Solar Millennium and others, are now coming into the U.S. market, is because there have been over recent years a number of very favorable policies in European markets that have really spurred activity.

In Spain, there is a feed-in tariff that is, essentially, a guaranteed something like 21 Euro cents per kilowatt hour that is paid to anybody who generates 50 megawatts, or less than 50 megawatts, and just, essentially, develops a project that goes into the nationalized grid in Spain.

So, of course, with that kind of an incentive, lots and lots of activity has been spurred in the Spanish marketplace. Other countries in Europe have similar measures, and the markets are growing rapidly in those countries.

Well, that has enabled companies to, like I just mentioned, to develop economies of scale, achieve technology advancements, and make the technology more affordable in the U.S. market.

As Mark pointed out earlier, we have the best solar resource on the planet. In all of Spain, Spain's solar resource would not have even shown up on Mark's map. And so, and I am quite serious, this is about 6.25, is that right, is kind of the high number that they strive for in Spain, compared to the 7.5 or eight kilowatt hours per square meter today that we have in this area in our home state.

So, the combination of a really desirable solar resource in the southwestern states, and, particularly, our home state, that and the credible developers who are not coming into the marketplace, I think create a really good recipe for expanded growth here.

Ms. Giffords. Okay, thank you.

And briefly, Ms. Rauluk.

Ms. Rauluk. I think it is important to remember, it is useful to have a feed in tariff or something that provides the extra value for the kilowatt hour that is produced from solar energy. That is a needed link in the marketplace. And, that is why Europe has really gone beyond what we have done in the U.S., because their value attributed to the kilowatt hours is significantly higher than it is in the United States.

But, I think we have to remember that who pays that, because that dollar amount for the extra value that you are paying for has to come from the rate payers, effectively, and there is really a limit to how much you can ask the rate payers to pay.

In Europe, it was a little bit easier. First of all, they have fundamentally higher electricity rates. So, if you are adding a penny to the kilowatt hour it is not a 10 or 20 percent increase, but it is, you know, less than that, and also the electric power industry is structured a little more simply.

So, my greatest concern is, I would love to see that the incentives available in the United States would be in excess of .20, .30 cents a kilowatt hour, you would get massive amounts of solar energy put in place. But, fundamentally, somebody has to pay, and, you know, how do we do that, and that is something we really need to think about.

Thank you.

Ms. Giffords. Thank you.

Mr. Hall.

WHY DOES SOLAR ENERGY NEED SO MUCH ASSISTANCE?

Mr. Hall. I guess to follow up, Ms. Rauluk, why does solar need so much assistance to be a viable source of energy?

Ms. Rauluk. Well, sir, there is a couple of reasons, one of which is that you are paying for all of your fuel up front, so that is the number one thing. But, fundamentally, the solar energy industry is relatively small compared to other energy industries and scale really, really is important.

So, the incentive structure, the way it has always been envisioned, and this is something that we have been doing in the U.S. for the last five to eight years, the incentive structure is to bring the scale of the industry and all of the manufacturing efficiencies, et cetera, et cetera, to bear on the problem and bring the cost down.

And, basically, we are realizing, though, efficiencies in both cost and performance, and this is a relatively medium-term incentive structure that we are talking about, and I would like to point out that there is not an energy system on the planet that has not been heavily incentivized because it is an important matter. We need to have reliable energy sources.

So, to do this for solar is really no different than what we have done for oil and gas, and I could go on into the list.

Mr. Hall. Well, I guess I was listening for you to say the reward would be great if we really could conquer this solar thrust, but you know the cost of solar energy can, if I was listening to the testimony right, can only come down and become competitive if the Federal Government, through carbon regulation, forces fossil fuels higher.

Is that what I am hearing? Is that your recommendation?

Ms. Rauluk. Actually, I think in some markets solar energy is competitive right now. If you have a market where peak power prices are in excess of .14–.15 cents a kilowatt hour, which they are in some markets, there are ways that you can put the installation together where it actually is fairly cost effective.

Mr. Hall. I think the rewards would really be great if and when we can conquer the problems with solar energy. It is unlimited, the reward would be unlimited. But, I see across this country a major war against fossil fuels at this time. I am from a fossil fuel state. Texas is one of ten states that produces energy for the other states. I see a thrust toward knocking out fossil fuels. If we knocked out fossil fuels, even in the next five years, these lights go out on us.

We get 60 percent of our energy from countries that don't like us. Our goal ought to be toward trying to lessen that percentage so we are not dependent on people that hate us and fly our airplanes into our buildings to kill our people. We need to really be addressing that, and solar can really help to do that, if we could find the money to put in there. But, I do not think we can find it by knocking down fossil fuels, when it is all we have now, and all we get from Saudi Arabia, 40 percent of our energy comes from them, is solely, totally, completely fossil fuels.

I think if we are going to declare war on something, we need to declare war through technology, finding cleaner fossil fuels and finding a way to do better while we seek solar. I am very fond of solar, and I think it has unlimited possibilities.

Mr. Mehos, in your testimony you indicate that without the continuation of the investment tax credit new capacity is going to be delayed by about 10 to 15 years. How quickly do you think the capacity could be developed if the tax credits were extended? I know you cannot say it is going to be six years, 10 days, and 45 hours, but just give me a good estimate of it.

Mr. Mehos. I think the best answer to that is to look at what Arizona Public Service and Abengoa are doing with the 280 megawatt project. That plant probably has a construction period of, oh, let us say around two years, not knowing that specifically. But, with the investment tax credit, if that begins to roll along, what we will see are those sizes of plants being built yearly, and probably multiples of those plants.

So, if I had to guess, I would say 500 megawatts to a gigawatt per year, even in the near-term, and let us say the near-term is in that five to 10, once we get past this four-year threshold, then every year after that 500 megawatts to a gigawatt or more per year.

Mr. Hall. In your testimony you also indicate that without the tax credit, solar would not be very competitive with conventional energy plants for quite a while.

Should the tax credit not be available, what do you envision the cost to consumers would be compared to conventional sources of energy that we have right now?

Mr. Mehos. Without the investment tax credit, using concentrating solar power as a proxy for solar, and it is probably the least cost technology of those at this point, the conventional cost from our concentrating solar power plant is probably on the order of .17 or.18 cents, let us say, a kilowatt hour, without the investment tax credit.

If we compare that to conventional technology in the intermediate load markets, that is about a 50 percent capacity factor.

For a combined cycle plant, you are probably looking at .10 or .11 cents per kilowatt hour. So, we are looking at, roughly, 50 or 70 percent higher.

ENVIRONMENTAL EFFECTS OF USING SOLAR POWER

Mr. Hall. Ms. Lockwood, I think my time is about up. In your testimony, you indicate that since the Federal Government is the largest landowner in the United States we should study the use of available land resources for CSP development.

Now, what are the effects to the environment from the use of solar power?

Ms. Lockwood. Congressman Hall, the effects to the environment depend on the particular location that you are in. The most obvious and clear impact is the amount of land that is consumed.

We are fortunate here in the desert southwest that we have large tracts of unused land that is very well suited for solar power. We certainly have to consider homes of exotic species and other types of environmental impacts when you are looking at siting solar, but we have large tracts of land that are very well suited for this technology today.

Mr. Hall. The use of all of these aides to the pursuit of solar power, you mentioned that, and I would ask you, would you support the use of closed military bases, you know, BRAC closes a lot of bases around the country, I think every 10 years, and this is just a suggestion to you to be thinking about, because you seem to be championing that. Do you support the use of closed military bases for development of large-scale solar projects?

Ms. Lockwood. Congressman Hall, I believe that is a perfectly valid opportunity for putting that land to use.

Mr. Hall. I know BRAC has a provision for refineries being built where you lose a BRAC. I don't know what your State law is, what BRAC closed down for the State of New Mexico, but we lost several in Texas, and they were cut down all over the United States.

Ms. Giffords. Mr. Hall, we do not allow them to close in southern Arizona either, and you realize that.

Mr. Hall. Yes, I know you would not allow that. You get the pitch forks out.

But, we put a provision in there. EPA is the major problem to getting permits to do things, and we had a provision in there at one time that if we made a request to EPA and they did not deny it in 30 days it was granted. And, I know you would like that, wouldn't you? We liked that. I don't think it made it through the Senate—very little gets through the Senate nowadays. But that is a good way to get refineries. Refineries are the reason gas is going higher; there are no refineries. Companies like Exxon and others do not want to put money into it, it takes 29 or 30 years to get their money back.

EPA would do nothing and we cannot appeal from nothing. So, we would rather have them turn us down in 30 days, and then we can appeal it, or grant it in 30 days and you go on with it. That is something that you might think about as you support the use of closed military bases, because I agree with you on that.

I think my time is up. I wish I had more.

Ms. Giffords. Thank you, Mr. Hall.

Vice Chairman Lipinski.

Increasing the Efficiency of Solar Cells

Mr. Lipinski. Thank you, Chairwoman.

First of all, I want to note that it is good to see, we have almost as many engineers on this panel as we have in the House of Representatives. I am one of the few engineers in the House. I also notice we have two Stanford alums here—I have an engineering degree from Stanford—and one from Berkeley unfortunately.

But, I want to start out in a little different direction in terms of the technology involved right now, and where we are going with that. I co-hosted a nanotechnology showcase a couple weeks ago in Washington to see some of the new products that are coming out using nano technology. And, I know at the University of Illinois they have done some work and found that by placing silicon nano particles onto silicon solar cells they can increase the power by about 60 percent and increase the life of the cell.

Where is this work right now, in terms of improving PV cells, and how much of a difference is that going to make in the near future, near to short-term, to mid-range future, in terms of how efficient solar energy is? Whoever wants to tackle that one.

Mr. Hansen.

Mr. Hansen. I will try to take that one, thank you.

TEP did invest in a manufacturing company, Global Solar, and I was involved in the technology looking for that, so I have some background in photovoltaics. In fact, I have a preference for photovoltaics as opposed to concentrating solar power.

You know, back in 1957, when silicon and gallium arsenide were used to develop photovoltaics, they were the predominant metals, if you will, for the use at that time. Since then, we now have those and efficiencies have improved from the less than one percent in 1957 to the 15 to 22 percent for silicon-based. Gallium arsenide-based are now almost at 40 percent, and we do have some materials over 40 percent efficiency. The—are more for silicon and sigs of copper—desalinate have all improved their efficiencies over the last decade. All the work with global solar weighs efficiency for long-scale production runs from two percent to over 11 percent. So, all of these technologies are improving.

I think we need to be careful not to focus on the efficiency when we talk about utility-scale, but to focus on cost per installed kilowatt. We have a lot of land in Arizona. I do not live in Tucson, I live in Apache County in the northeastern part of Arizona, when I travel from my home to our coal-fired power plant, and our solar plant, we have about 4.5 megawatts of solar photovoltaics there, I pass by approximately 150 square miles of land that has about, as I like to say, eight bushes and one house on it, and there is a lot of room, it is fairly flat land. We have space in Arizona.

Efficiency deals with space. Cost is going to be the driver. We need to be improving efficiency, but we need to keep our eye and our focus on reducing the costs. The nano technologies that are now being developed, and some of the organic dye technologies as well, and some of the more advanced thin films, do show promise to be able to reduce the cost of the photovoltaics to dramatically lower numbers, talking in numbers that are less than a dollar per watt, whereas, conventional technologies today are typically at the module level in the neighborhood of $4 to $5 a watt, some as low as $3.50. This is what will drive the cost, and that is really the issue on photovoltaics, is the cost. We need to bring the cost down.

Ms. Rauluk. I would like to mention a couple of things. First of all, and I am not an expert in all of the technology improvements and innovations that are in the pipeline right now, but just from my discussions and work with my colleagues at the University of Arizona and Raytheon Missile Systems, these folks have viable technologies that are, I would call it, in the final stages of R&D. So, there are some very exciting and interesting things in the PV, the concentrating PV, or CPV, and the highly-concentrated PV area that are coming out of the laboratory.

Now, there are incremental changes, incremental improvements that are happening with PV, and when people look at, well, what is the efficiency and the cost, and how is this all working out, no one is thinking about, well, what is coming out of the labs, because the commercialization process is a difficult process with some uncertainties attached to it.

But, let me just point out that, you know, a lot of people say, well, let us not do anything until the technology is really great, and then we will just go and one of the things that supports the technology in the labs right now is the recognition of a marketplace existing and viable. So, when these technologies are coming out of the lab, which I expect they will within the next three years, they will need to get financed by venture capital and then second-stage financing, et cetera, et cetera, and people are going to look at that and say, well, where is the market? So, even if we have a good technology, it is important that we do not wait until we get these things coming out of the labs, but that we have a systematic and reliable investment plan for the future.

Mr. Lipinski. Thank you. I see a red light, and I will yield back. If we have a second round, I will have another question.

Ms. Giffords. Okay, thank you.

Mr. Lipinski. Thank you.

Ms. Giffords. Mr. Matheson.

ACCELERATED TECHNOLOGY INNOVATION

Mr. Matheson. Well, thank you.

I am not an engineer. I am, my background is in finance, but I used to be an independent power developer, and developed co-generation facilities. And, now I sit in Congress, where we try to come up with public policy ideas to help foster these new technologies.

And, it seems to me that we need to be looking at this on a couple of different paths at the same time. I have heard all the witnesses talk about the need for, lack of a better term, federal subsidies to help create large-scale commercial applications of technology.

What I am curious about is another path that we also ought to be talking about, I think, and that is the notion of how do we get these technologies to be more efficient so that they are commercially viable, perhaps, with less subsidies, or, perhaps, with no subsidies.

And, Mr. Mehos, your last slide, at the end of your presentation, the kick-starting utility-scale solar slide, you mentioned that there is an exercise for DOE that is estimated at $50 million a year to achieve the accelerated goals. And, that was what I was wondering about, is that program. It sounds to me, you can confirm this for me, but DOE has identified a path to help accelerate this technological innovation, and I think as Members of the Science Committee that is, obviously, something we have great interest in as well.

So, can you share with us a little bit about what that accelerated effort entails?

Mr. Mehos. Sure. The accelerated effort, as I briefly mentioned, in along two paths. It is continued technology development on the specific technologies, trying to achieve higher temperatures, and I will describe that in a second.

The two technologies in the concentrating solar power program that achieve these higher temperatures are the line focused parabolic trough technologies and the more point-focused central receiver technologies, as well as the point-focused disturling technologies. But, of those first two, the parabolic trough and the central receiver technologies, going to higher temperatures achieves a couple of things. One, it allows you to operate your cycle at higher efficiencies, that does decrease the levelized cost of energy. As importantly, or maybe even more important, as you go to the higher temperatures we are dealing with high amounts of thermal storage materials having a higher delta T difference between your hot temperature and your cold temperature to work with, significantly decreases the amount of thermal storage that you are working with, and that, in itself, also decreases the cost of thermal storage, or of the levelized cost of energy.

So, those are two of the higher pathways, higher temperature pathways, that lower your cost.

We are working on a number of other avenues. We are looking at higher temperature materials, higher reflectivity materials, better absorbing, less emitting materials. We had a study some time ago now, I think back around 2002 with Sergeant Inlundy, basically, identified three mechanisms for reducing the cost of solar power, none of which were power peaked. The first one is the research I described. That results in about a 30 percent reduction in the cost of electricity. The second one is just increasing the size of your plant, the type of work that APS and Abengoa are working on, going from smaller to 280 megawatts. And, the last one is learning, it is deployment, and the more you deploy these technologies, actually, this does get into policy, then the lower the cost of the technology over time.

Financing Technology Development

Mr. Matheson. And, is it fair to say that, I mean, obviously, one of the variables that helps this process go is if Congress appropriates the funds so that this effort can happen. Dollars, you know, money is part of making this technology develop.

Mr. Mehos. Yes.

Mr. Matheson. Are there other policy options that also need to be considered or adjusted that Congress hasn't done that could help facilitate the development of these technologies?

Mr. Mehos. Yes, I believe so. I think in project finance one of the key issues is risk. When we start talking about $1 to $2 billion projects, I mean, the risk associated with that is relatively high. The parabolic trough technology is actually fairly low risk, but still project finance can be an issue.

And, as you look toward some of these higher temperature technologies, like the central receiver, or the disturling technology as an example, the policy of loan guarantees, or federal loan guarantees, comes to play there.

Mr. Matheson. Okay, that is helpful.

One other slide that I wanted to ask you about. You showed the difference of whether the 30 percent ITC is extended or not, and the roll out of solar technology. It is probably impossible for you to estimate, because we do not have a policy in place yet in this country, but did you consider if the cap on trade program is put in, and there is a price associated with the carbon, how that would affect the curve when you are developing those drafts?

Mr. Mehos. No, that's a good question, and we have considered it, we do not have the ability to model that yet.

Mr. Matheson. Yes.

Mr. Mehos. That is one of our outcomes this year, we will be able to model those types of systems.

Mr. Matheson. Mr. Hansen, this is a little off target, but you mentioned smart grid when you were talking, and, you know, the Congress just passed smart grid legislation in the energy bill that passed last year. Do you feel like the legislation that Congress passed was helpful for smart grid? Are there other things we should be doing beyond what was in that legislation, or do you have any thoughts on that?

Mr. Hansen. Good question. What you passed is very helpful. It is a good start. What we need now is additional, kind of what Valerie just alluded to, we need scale. That is up to the utilities.

Over time, over working with our individual State regulatory entities, we will be able to get recovery for those additional costs.

Mr. Matheson. Right.

Mr. Hansen. I think the Federal Government has stepped up to the plate and given us the tools that we need from the federal level. I think we now need State level to step up to the plate.

So, I appreciate the efforts you had last year. Thank you.

Mr. Matheson. Great. Thanks.

Madam Chair, I see my time is expired.

Ms. Giffords. Thank you, Mr. Matheson.

Mr. Mitchell.

LAND USAGE FOR SOLAR POWER

Mr. Mitchell. Thank you.

One of the things that was mentioned earlier was that to be successful large-scale solar facilities need land, and the Solana project is three square miles. Is there any possibility that we can be able to, with technology, lessen the need of land? You know, also mentioned was that the reason that there was a market in the southwest for solar was because we had the land, and that we had high renewable portfolio standards.

One of the fights we had over this last energy bill was over the national portfolio standards, and it was knocked out mainly because there were states that said they really didn't have the land, or they didn't have the sunshine, and as a result the portfolio standards are really out here in the southwest.

Is there anything we can do maybe to help establish a higher portfolio standard nationwide at the same time maybe not use as much land as we are going to be able to use here?

Anybody.

Mr. Hansen. If I may, again, going back to the question asked earlier about efficiency, improving the efficiency, raising the efficiency of photovoltaics, raising the temperature of the collection on concentrating solar, will reduce the amount of land that is required.

Every state has the ability to produce some level of solar. There was a study done a few years back by Black and Beech on the State of Pennsylvania. I went to school in the State of Pennsylvania, I grew up in New Jersey. I don't remember seeing the sun a whole lot of the time, but, quite frankly, the result of that study indicated that the only renewable resource that could meet all of the energy needs of Pennsylvania was solar. So, every state does have the ability to put in solar.

The roof tops are available, without having to use any land. There is a wide range of solar technologies available. TEP's experience is with about 12 of those at this present time, and what we have found is that different technologies have advantages in different climatic zones, as I said in my written testimony. All of the technologies need to be developed, so that we, as a United States, have a portfolio of opportunities. We, as utilities, can pick and choose among these different technologies as to what is most appropriate for our state.

I am not going to weigh in on the issue of a national renewable energy standard. I think the State of Arizona has stepped up to the plate and done an excellent job leadership-wise in developing one that is appropriate for Arizona. But, I do think that over time, with the federal level support, with the national labs, and with universities, the funding can be provided to improve the efficiency, to improve the overall storage capability, for energy in the future that is going to drive things like the solar, Grand Solar Plan and these other technologies to economic fruition, so that they will, in fact, become economically compatible with coal and natural gas.

Ms. Rauluk. One of the reasons why distributed generation has value is that you don't need really large contiguous pieces of land. And, in fact, you can, Mr. Hansen already mentioned, you can put photovoltaics on roof tops, but you can also put photovoltaics in smaller pieces of land and every community has pieces of land that may be old industrial sites, next door to an old industrial site, maybe it is a buffer for an airport, whatever, that are not, you know, several square miles, but are maybe a half a square mile, or even less than that, and you can use that land for a distributed application, because it can scale down to that.

So, I think that the amount of contiguous land doesn't really constrain us when we are talking about a distributed format, and there are plenty of opportunities to do that.

Price of "Green" Power

Mr. Mitchell. One last question, maybe this is Ms. Lockwood.

You know, some people pay an extra little premium to encourage green production of power. Is there any way that anybody is going to take advantage of that once Solana comes on line, or is it all just going to the grid and everybody still pays the same price?

Ms. Lockwood. Congressman Mitchell, we very much believe in the power of our customers to drive the policy and our resource choices. So, absolutely. Solana is several years away, but we do envision a way that our customers can choose to pay a small premium and have all or part of their energy served by Solana.

Ms. Giffords. We only have a few more minutes left, and since I am going to afterwards ask our witnesses to come up, I am going to defer my questions, maybe just one additional question from each of the Members before you have to leave.

Utility-Scale versus Distributed Generation

Mr. Lipinski. Yes, I will lead off here. I just want to know, we are in heaviest discussion here about utility-scale mostly, but also distributed generation. Is there any conflict or tension between the two, in terms of one obviating the need for the other? I just want to throw that out there and see some smiles on faces about this. It must be something that you deal with regularly.

Mr. Hansen. If I may, they can be compatible with each other. TEP's studies have indicated that for us to produce 10 percent of our annual energy from solar we need about 610, 620 megawatts of solar installed.

If every home in Tucson were to have about a three kw system, which is realistic in size, that would give us about that 600 megawatts of solar.

It is at about that point when the energy storage becomes a critical component, if we are going to move solar beyond that 10 percent of our annual energy. That is why TEP has always been trying to develop a balance of distributed generation as well as utility-scale.

Even if every home in Tucson were to install nine kilowatts of photovoltaics, that is 30 percent of our energy. The other 70 has to come from someplace, and we are proposing, at least I am proposing in the long-term, that that comes from the utility-scale solar, such as the 280 megawatt system that Abengoa is planning to put in. But, it could also be from photovoltaics.

In the long-term, the two systems have to mesh, and the glue that makes them mesh is the storage. Even distributed generation without utility-scale solar is going to require some level of storage to even out the day to night intermittencies.

The other part of the puzzle that is hard to understand for some people who have not lived in the southwest is that most of the solar energy is actually produced in the springtime in Arizona, away from our monsoon storms. But, of course, most of the consumption is in the summertime, so we have to shift about three months worth of solar energy into the summertime. It actually works out to be approximately 10 percent.

One other factor I think that needs to be considered for the future, is plug-in hybrid electric vehicles. Our calculations indicate that with an additional 10 percent of energy per year we could provide all of the energy that is needed for all of the passenger vehicles in the City of Tucson. So, that does not include heavy trucks, and airplanes, and locomotives, but your normal passenger vehicles. That, again, could be derived from solar, and could provide an additional opportunity for storage if we go back to that smart grid development and how to integrate them as part of this storage philosophy.

Mr. Lipinski. Ms. Rauluk.

Ms. Rauluk. I have alluded to this in my spoken comments, and I have a more detailed explanation of it in my written comments, but there is a fundamental conflict right now in the way in which we contractually do these things between distributed generation and the utilities revenues, because the distributed generation is on the customer's site, and they are effectively purchasing less energy from the utility by generating their own energy.

And then the question arises, well, how do you assure the preservation of utility revenues and the assets that they support, because this is not about getting rid of the utilities or getting rid of the fossil fuel generation by any means.

So, there are things we need to do contractually and from a regulatory point of view, and the industry is well into the phase of doing that and creating the mechanisms that, basically, do not conflict, do not have the utility having a natural and inherent animosity towards distributed generation, but that it is a part of the whole system and is valuable for the whole system.

Mr. Lipinski. Ms. Lockwood.

Ms. Lockwood. Congressman Lipinski, I do not know that I disagree with anything Mr. Hansen or Ms. Rauluk said, but for us in Arizona, in particular, for APS, it is about growth, and we are growing so fast, our energy consumption is also growing so fast, that we need all resources to meet our energy needs into the future.

We believe that both are required to get where we need to go, and do not believe there is a fundamental conflict. Now, there is some theory that there is only so much subsidy or incentive to go around, and I think that is where a lot of the debate comes in. Does it go to large scale, or does it go to distributed? And, that is a healthy debate. That is something that we need to be talking about. There are different economics when you look at those different sides of the issue.

For utility scale, we very much look at it in comparison to the other resources that we have. Even without carbon today, large scale is getting—large-scale CSP, solar thermal—is pretty competitive. Our Solana plant is about a 20 percent premium over our conventional resources, what we would have expected to pay for fossil fuel resources into the future for that project.

On the distributed side, you look at it not what you pay for other large-scale generation, but you look at what the customer is paying and the offsets for that customer, and how that works within your rate structure also.

So, from our perspective we need them all, and we need to make sure that we are looking at policies that facilitate them all in the appropriate way and the appropriate manner, also considering the economics and how the impact to the rate payer.

Mr. Lipinski. Thank you. I thank all the witnesses for their testimony. It was extremely helpful today, and thanks to Congresswoman Giffords for bringing this together.

Ms. Giffords. Thank you.

Mr. Matheson.

Compressed Air Storage and Greenhouse Gas Emissions

Mr. Matheson. Why, I'm nervous about going over the deadlines that Madam Chair set. Let me ask one real quick question. I have got to chance this.

I was reading about the Grand Plan, and, you know, one of the great benefits of solar in a world where we are concerned about climate change and global warming is that we move away from fossil fuels.

But, I did note that in the energy storage component of the Grand Plan, we are going to use compressed air, there would be some degree of natural gas used. Do you have a sense of what that means in terms of greenhouse gas emission?

Mr. Hansen. The use of the natural gas for the reheat on the turbine, and this is, before I say that, this is a technology that can be changed. You can make turbines that do not have to have this natural gas input.

Mr. Matheson. Oh, okay.

Mr. Hansen. Alternatively, you could be using biomass or some other type of fuel, bio-diesel, et cetera, to do it, but it is approximately one-sixth of the input that otherwise would be required from natural gas or coal under a normal conventional technology.

Mr. Matheson. Okay, thank you.

That is it.

Ms. Giffords. Thank you.

Mr. Mitchell. I don't have a question. I would just like to thank everyone, because it was very informative, and not only the written material but your testimony.

So, thank you all very much.

Ms. Giffords. Well, before we bring this hearing to a close, I want to again thank our witnesses for the generous time and for really a very, very interesting discussion.

The Science Committee is, I believe, the bipartisan Committee in Congress, and we have been able to do many things just in the last a little over a year, that I think this country would be very proud to know, if they had a chance to hear about it.

Unfortunately, usually when we have committees the bells ring frequently, so we will have wonderful testimony, and then we will have to jump up, run over and vote and come back. So, what a luxury to actually have a chance to really focus on the information that you have presented before us today.

The record is going to remain open for additional statements from the Members and answers to any follow-up questions the Committee may ask of our witnesses.

I would also like to thank the bipartisan Science and Technology Committee staff for being here, for coming out from Washington to help conduct this hearing. Also, members from my staff, Tamarack Little, Wyatt King, Jacqueline Jackson, are just a couple that have worked so hard to bring this committee here to southern Arizona as well.

I want to thank the solar experts, and there are many in the room today, and, hopefully, we will have a chance to hear from you in a couple of minutes, because we are going to ask our witnesses to come forward and to answer questions from the general public as well.

But, to the public, thank you for caring so much about the future of the southwest, the future of our country, and the future of how we can take this tremendous potential, harness it, and turn it into some real energy.

So, with that, the witnesses are excused and the hearing is adjourned. Thank you.

[Whereupon, at 2:37 p.m., the Subcommittee was adjourned.]

APPENDIX 1: ANSWERS TO POST-HEARING QUESTIONS

Responses by Thomas N. Hansen, Vice President, Environmental Services, Conservation and Renewable Energy, Tucson Electric Power

Questions Submitted by Representative Adrian Smith

Q1. You spoke about the need for transmission and storage for solar energy. When built, do you envision this infrastructure would be available to other renewable energy technologies, such as wind, an important Nebraska resource?

A1. Most definitely. I envision a large number of storage facilities distributed around the country. The storage should ideally be located locally with respect to where the electric consumers are located, effectively on a regional basis, with one storage facility serving at most five million people. Locating underground energy storage locally provides the optimum solution for energy security for the residents in that area. The nationwide electrical transmission system would allow for the movement of wind energy produced in Nebraska to be stored in Nebraska or California or Maine. Just as our Interstate Highway System enables goods produced in one area to be efficiently delivered in another region, the interstate transmission system would efficiently enable solar, hydro, wind, geothermal, tidal, current and biomass energy to be moved around the Nation in a controlled manner to maximize efficiency of production and delivery. The combination of properly sized local storage with national transmission would allow management of the overall system to minimize congestion.

Q2. As a Member of the House Science and Technology Committee, I have a keen interest in NASA and the space program. Could solar energy collected in space be a viable source of energy for the U.S.? What are the benefits and challenges of this technology?

A2. This concept has been discussed for decades and is, in my opinion, technically viable although some components of the technology need to be improved in terms of reliability and efficiency. One big advantage includes better solar intensity above the atmosphere and 24/7 solar production potentially without day/night cycles or clouds to block the sun. This results in much higher specific energy production per unit area of solar collector reducing the size of the solar collector by 80 percent or more. Building a multi square-mile solar collector in space will be challenging in terms of providing sufficient resources of material and people to a synchronous orbit. Wireless transmission of energy from space to Earth would require very accurate targeting systems and dedication of a few square miles of receivers and buffer zone on the ground to convert the beamed energy back into grid power at high efficiency. However, a large single energy receiver at the Earth end of a space solar energy collection system could be at risk from an act of terrorism, while multiple receiver zones would present more risk of component failure and resulting repair, in addition to an increase in initial cost. It also may be challenging to convince people that it is safe to live near a receiver zone. A space bound energy collector would be at greater risk of damage from collision with meteorites without protection from the atmosphere. Maintaining optimal orbital geometry to enable a space bound solar collector to keep sight of the sun at all times while also keeping its energy beam to Earth on target will be technically challenging, but not impossible. Interestingly, given that the space located solar collector would produce energy at a constant rate, energy storage would still be required to balance the constant energy input with a variable energy demand. The national transmission system would be required to allow for delivery of the space produced energy from a single, or small number of multiple, Earth side satellite energy receivers to all U.S. energy consumers. Some larger questions that still need to be answered are economic: What is the total cost of such a space located solar energy production system? What is the energy balance —will it take more energy to place the energy system in orbit and maintain it than the system will produce over its lifetime? Will more valuable jobs be created for Americans with a space bound energy collection system or a terrestrial located energy collection system? Both the space bound and terrestrial solar energy collection concepts deserve further consideration, although terrestrial solar energy collection technologies are fully developed and available commercially today.

Responses by Valerie Rauluk, Founder and CEO, Venture Catalyst Inc.

Questions Submitted by Representative Adrian Smith

Q1a. You spoke about distributed generation systems, in which smaller generation systems (rooftop units and 10–50 acre land units) spread electricity generation over "unused real estate" and reduce risk of large scale power outages. Would the Federal Investment Tax Credit provide incentive for individuals to install smaller generation systems (e.g., rooftop units) and to become a part of a distributed generation network? If not, how could the Federal Government encourage this type of development?

A1a. Yes, the FTC does provide incentives for individuals interested in a distributed generation application and participating in a network.

Q1b.How would you envision the development of distributed generation systems? Who will pay to connect these smaller generation systems into a cohesive network?

A1b. In addition to extending the FTC set to expire at the end of this year, the Federal Government could further encourage such installations by setting distributed generation requirements for utilities nation-wide and to encourage incentives and research and development (especially commercialization R&D) for distributed systems and the intelligent controls and storage options that increase a DG network's value and resiliency.

Q2a. As a Member of the House Science and Technology Committee, I have a keen interest in NASA and the space program. Could solar energy collected in space be a viable source of energy for the U.S.?

A2a. Theoretically, it could be subject to resolving certain technological challenges. However, in the near-term, there are many cost-effective opportunities for harvesting solar energy on the surface of the planet.

Q2b. What are the benefits and challenges of this technology?

A2b. The chief challenge is delivering the collected solar energy to where people can use it, in electric power parlance, the ''load.'' The cost of delivering to the load from remote locations on Earth is one of the fundamental challenges and costs and is why solar energy in a distributed format is more beneficial than central station applications. Energy generation from space would create even greater costs and challenges. However, there may be some benefits to doing so. I have not reviewed the literature concerning this option and cannot offer any insights into the benefits.

Responses by Joseph Kastner, Vice President of Implementation and Operations, MMA Renewable Ventures LLC

Questions Submitted by Representative Adrian Smith

Q1. As a Member of the House Science and Technology Committee, I have a keen interest in NASA and the space program. Could solar energy collected in space be a viable source of energy for the U.S.? What are the benefits and challenges of this technology?

A1. It is my understanding that the DOE studied the collection of solar energy with photovoltaic panels is space several decades ago. Some of the large hurdles for this idea include providing a safe, efficient means for transmitting the electricity to a terrestrial collection point (the DOE contemplated using microwaves) and the mobilization of a large-scale construction project in space (to make it worthwhile the array would be many times larger than the International Space Station). Such a large array would also be quite susceptible to space debris.

End Notes

[1] The assumptions here are:

- A state economy (GSP) of $250B (a median range across states);

- Only direct jobs—no manufacturing or other indirect jobs are considered here;
- Investment represents only direct capital associated with the plant and assets;
- GSP increase includes indirect and induced effects.

CHAPTER SOURCES

The following chapters have been previously published:

Chapter 1 – This is an edited, excerpted and augmented edition of a Solar America Cities Publication, dated April 2008.

Chapter 2 – This is an edited, excerpted and augmented edition of a United States Department of Energy, Solar America Initiative, Solar Energy Technologies Program publication, Draft dated February 5, 2007.

Chapter 3 – This is an edited, excerpted and augmented edition of a United States Department of Energy, Solar Energy Grid Integration Systems "SEGIS" publication, Program Concept Paper dated October 2007.

Chapter 4 – These remarks were delivered as statements during a hearing before the committee on Energy and Natural Resources, U.S. Senate, dated July 2, 2008. Hon. Jeff Bingham, Chairman, presiding.

Chapter 5 – These remarks were delivered as statements during a field hearing, before the subcommittee on Energy and Environment, U.S. House of Representatives, dated March 17, 2008. Hon. Gabrielle Giffords, Vice Chairman of the Subcommittee, presiding.

INDEX

A

B

C

D

E

F

G

H

I

J

K

L

M

N

O

P

Q

R

S

V

W

Y

Z